Quorum Superhealing

Reclaim Nature's Quorum Nutrition™ and Timeless Wisdom to enhance your Body's Inner Physician

Paul Yanick, Jr. Ph.D, ND, CNC, CQM

Order this book online at www.trafford.com
or email orders@trafford.com

Most Trafford titles are also available at major online book retailers.

Note for Librarians: A cataloguing record for this book is available from Library and Archives Canada at www.collectionscanada.ca/amicus/index-e.html

Printed in Victoria, BC, Canada.

ISBN: 978-1-4269-1682-3 (soft)
ISBN: 978-1-4269-1681-6 (hard)

Library of Congress Control Number: 2009935541

Our mission is to efficiently provide the world's finest, most comprehensive book publishing service, enabling every author to experience success. To find out how to publish your book, your way, and have it available worldwide, visit us online at www.trafford.com

Trafford rev. 9/21/2009

www.trafford.com

North America & international
toll-free: 1 888 232 4444 (USA & Canada)
phone: 250 383 6864 • fax: 812 355 4082

Contents

Acknowledgments

There are many people to thank in seeing this book come to fruition: The numerous colleagues who let me try out new ideas, the friends who patiently listened to my brainstorms, the professors and others who led me to deeper knowledge, and the hundreds who offered encouragement and insights. Still, I must single out a precious few for special acknowledgment. To my wife, Bonnie Lee, for her support, unconditional love, and friendship, and for her organization and management skills and my daughter, Heather Rose and son, Thomas Scott, whose love and support helped me to accomplish this task. Most importantly, I want to give thanks to our loving and compassionate God, Jehovah, who set forth the laws of nature and gave us many healing foods and herbs. To S & K Enterprises at www.sksigns.com for beautiful cover logo. To all of you who have adopted the principles of *Quorum Nutrition* and made it your own, thank you.

An Important Note to Readers

This guidebook represents decades of scientific research and study in the fields of anatomy, physiology, biochemistry, quantum physics, pharmacology, endocrinology, neurology, nutrition, psychology, anti-aging medicine, and Quantum Medicine. In an effort to make it more accessible to the layperson, I have avoided extremely technical language and omitted thousands of arcane scientific and textbook references. Many of the concepts presented in this book are indisputable facts of life that follow the laws of nature; they are based on natural laws and empirical observations, sound reasoning, and common sense. This book also reflects more than three decades of clinical experience with laboratory-guided observations of many clinical disorders. **Ask your doctor before you implement any of the lifestyle changes advocated in this book. If you have a disease or any other type of health-related problem, you should first consult your physician before attempting to deal with it. This book is not meant to replace a medical examination, and its contents are not meant to diagnose medical conditions, interpret medical symptoms, or render medical advice. It should be used in consultation with your doctor to better understand health problems and their possible range of treatments.**

The purpose of this book is to increase nutritional awareness and to educate individuals regarding natural ways to improve their general health and well-being. All readers are encouraged to seek help from doctors who treat people as whole individuals with unique and distinct dietary, exercise, and nutritional needs. The best course of action is for the reader to use common sense and the information in this volume in consultation with a reliable physician to achieve a healthy, fulfilling lifestyle.

Quorum Nutrition, LLC 3905 E Russel Rd, Suite E, Las Vegas, NV 89120

FOREWORD

"Everyone has a doctor in him or her; we just have to help it in its work. The natural healing force within each one of us is the greatest force in getting well. Our food should be our medicine. Our medicine should be our food."

Hippocrates, M.D -Father of Western Medicine

Hippocrates, the Father of Western Medicine who was one of the most honored names in medicine, taught that the healing power of nature empowered the "doctor" within or inner physician. Despite these revered words of wisdom, science and medicine know little about the complex operations of the body's inner physician. Nor have science and medicine studied to any degree the inner physician to explain why placebos can heal the sick or why there has been a frightening progression in all illnesses since the birth of modern medicine. And, in natural or complimentary medicine, only the tiniest portions of the inner physician have been tapped.

Doctors of modern medicine neglect to explore the origins of this seemingly inexhaustible healing power, treating it like an imponderable, inaccessible puzzle whose operation is too complex to understand. It is almost as if they fear losing authority or control over us or the prominence that comes from being a doctor.

Yet, in today's toxic and stressful world, tapping into this inner healer may provide us with the only way to heal and de-stress our bodies.

Why do so many people struggle needlessly with health problems when we are bombarded with information daily on how to become healthy? I believe that most of us are kept from our tapping into our true healing potential because we are held captive by subtle and destructive myths, lies, and fallacies about our health. These myths are crippling our inner physician and drastically and negatively influencing the quality of our lives on a daily basis.

Upon close scrutiny, you'll see that these myths are at odds with common sense and the laws of nature. Once you understand the latent powers of your inner physician and explore its true potential, you'll become a believer and question things that most people never question. You'll gain a clear understanding about its critical link with nature, how it operates, and what is myth and what is useful in your quest for better health.

The spectrum of your inner physician's capabilities is enormous; its potential power is prodigious and extraordinary. Once you tap into its superhealing power, you'll realize that popular majority opinions about health rarely point to the truth.

It's human nature to stick with the things we are familiar with and the things we think we know. For some of you, it might be a challenge to make sense of new things because your frame of reference is distorted and not founded on the truth. Subtle lies find fertile ground in the elements of fear and greed that pave destructive paths of short-term rewards while disguising the long-term devastations to our health.

In the face of a broken paradigm of healthcare, Americans are seeking greater control of their health and are reclaiming the traditions of our ancestors. With this expanded awareness, more are focusing upon how to awaken and activate the innate powers of healing. As health care costs increase and become inefficient at keeping us healthy in a toxic world, greater responsibility is being—and will continue to be—placed on each of us to stay healthy and prevent disease.

Most of us have grown up with a "fix it" state of mind when it comes to our health. We don't look deep within ourselves at our inner physician to find out why we are sick. We crave instant gratification and instant energy, finding it easier to block symptoms than to address the true cause of our aliments. We don't take care of ourselves as we should. We've forgotten and foolishly locked our powerful inner physician in a vault deep within our brain.

Our brain is hardwired for health. Even though living in today's polluted world is hazardous to our health, the physical hazards to health and life need not be so punishing. Despite the endless stream of toxicants that disrupt the bioelectric power of the inner physician, new solutions are now available to stop the widespread affliction of neuro-toxicity and malnourishment.

But to work properly, the brain needs to connect us more fully with the quorum cycles and polarities of nature. The quorum language of cells is being disrupted by our polluted world. We have lost contact with nature's quorum cycles of healing.

Your health depends on the ability of cells to talk to one another via what scientists call "quorum sensing" and the nourishment that you give your cells. But, generation after generation have slowly sculpted us to be more toxic and nutrient-depleted at the cell level. We are distressed and we must allow our distress to motivate a self-examination of our inner physician.

The future of our victory over disease depends on going back to nature or to reestablish a quorum of all cells in the body. Even though using nature's remedies will raise the eyebrows of the dominant medicine-profit system, rest assured that much of what ancient medical practitioners knew and prescribed from nature has been validated by modern science.

There are two general types of cells in the body: human and commensal-probiotic cells. Virtually every bodily process that both protects and heals us involves a quorum communication and interaction between human and commensal cells. This quorum language controls everything from wound healing to cancer cell destruction.

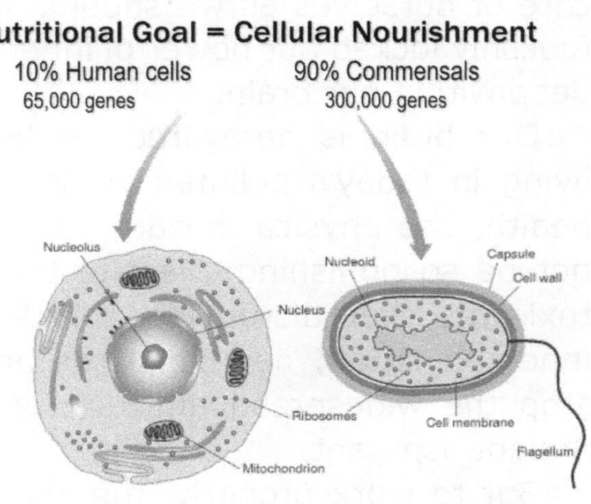

The term probiotic, which means "in support of life," has been misused in today's marketplace. Thus, in operational terms, I call probiotics *commensal cells* that occupy 90-95 percent of our body's total cell population. These miracle healing cells are the key to health and vitality and the primary source of quorum communication in the entire body.

Commensals provide the main ingredient for this language in the form of special nutrients, called *Quorum Nutrition*™. *Quorum Nutrition*™ keeps our hormones in balance, fights off disease invaders, enables the blood to clot, and gives neurons the kind of nourishment that can restore the full operational complexity of our inner physician. Today there are more than 5,000 research papers on commensals published worldwide and the key ones are cited in the Appendix if you want to learn more about the science behind commensals.

Food is one of the most powerful ways to engage with nature. But shelf life deterioration, soil depletion, processing and toxic chemical packaging have nearly wiped out our cellular nourishment. To make matters worse, stress and the constant onslaught of chemical pollutants have literally turned off our digestive physiology and poisoned our commensal cells.

Food, water, sunshine, and air remind us of our critical dependence on nature. Thus, to get well, we have to tune out all the nutritional advice that's out there—all the health claims, all the fad diets, and all of the synthetic, inorganic and milligram-dosed supplements and symptom-chasing games that are widely prevalent.

When you think about it, antibiotics taken at any one time in your life have killed off the majority of your commensal cells. This is in direct opposition to nature's recipe for health. To be healthy and engage our inner physician t the fullest level, we need ten percent human cells and ninety percent commensal cells.

Since man declared war on microbes with the advent of the pharmaceutical revolution in the 1920's, we have diminished quorum communication dramatically. Add to that all the natural anti-infectives (neem, oregano, goldenseal, garlic-derived allicin and silver which destroy commensal cells), and you can begin to see why there is so much suffering and premature death in the world.

To reconnect with the power of nature, we have to see ourselves as living ecosystems and put back commensal microflora to achieve symbiosis or cellular balance in our bodies. But, before we do that, we have the put back the quorum nutrients needed to jump-start commensal cells and allow them to find a permanent home in our gut.

The more quorum nutrition we can directly supply to the body, the faster we can rebalance our immune networks, drive our energy pathways, sweep away dangerous pollutants and carcinogens, and control the sophisticated cellular and neural reactions needed by the inner physician to heal our bodies.

When given the right pre-digested quorum nourishment, the inner physician has an enormous capacity to correct and heal the body. Commensal cells manufacture compounds that are ten times stronger than stem cells in regenerating the body. They are nutrient factories, pumping out tons of quorum nutrition to enhance tissue repair (wounds, sores, periodontal disorders, etc) and help us discharge fat-storing toxins called xenoestrogens that are responsible for the current belly fat epidemic.

Could the rapid increase in cancer and all diseases we see today be nothing more than commensal cellucide and nutrient deficiencies? It only stands to reason that we have to RESTORE normal and natural physiology or the original framework needed for quorum cell-to-cell communication used by nature in the *Quorum Symbiotic Cycle*.

Nature's *Quorum Symbiotic Cycle* allows a mutually beneficial relationship to occur between the sun, plant cells and soil bacteria. Soil bacteria feed the plant, and the plant feeds the soil bacteria. And, the same cycle occurs in the body. Commensal cells feed human cells while human cells in control of digestion, feed commensal cells.

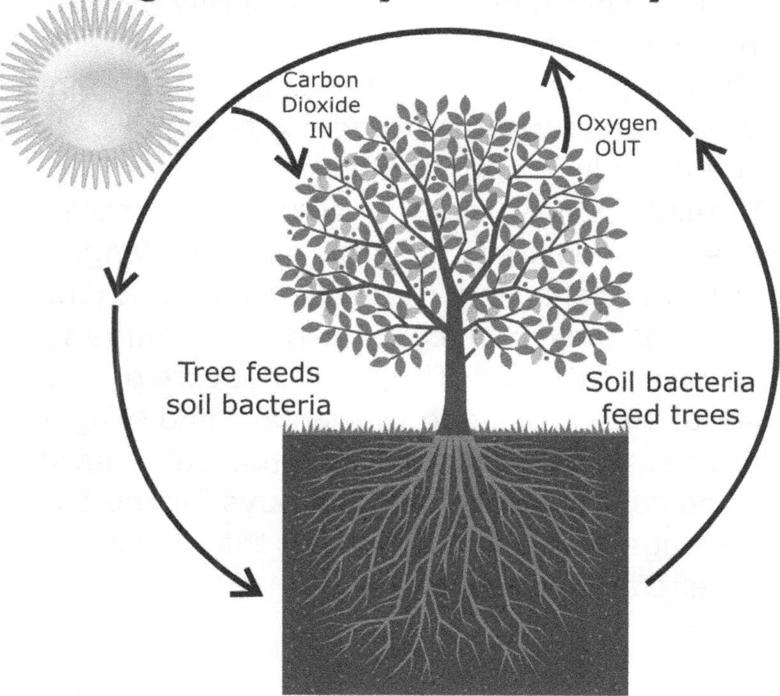

The energy in sunlight is converted into food by plants and soil-based bacteria. Photosynthetic organisms have large protein complexes that absorb light and convert it to usable cellular energy. In the quorum framework, each complex has a large, dense scaffold of proteins attached to hundreds of other nutrients. It is this kind of quorum nutrition that is the driving force behind our inner physician.

The more scientists learn about virtually every disease, the more they realize that the malfunctioning immune system and its commensal cell army is the real and silent cause of everything from heart disease to obesity.

Our cell-to-cell quorum communication is the ultimate connection we have to nature's healing powers, and it needs quorum nutrition to function. But since commensal cells are easily poisoned by man-made chemicals and pharmaceuticals and wiped out by antibiotics taken in the past, I had to invent a new class of nourishment and a way to put back what nature intended in our bodies.

Virtually every change within our multicellular bodies is mediated by the quorum language of commensal cells and the millions of nutrients and compounds they produce to keep our immune systems healthy and balanced. They lie at the core of your cell's ability to stop prolonged, unwanted, and exaggerated inflammation. However, with commensal cellucide, the immune cells mistake the good guys for the bad guys and engage in cell-to-cell combat that causes all sorts of degenerative disease.

Quorum Symbiotic Cycle **The Dysbiotic Cycle**

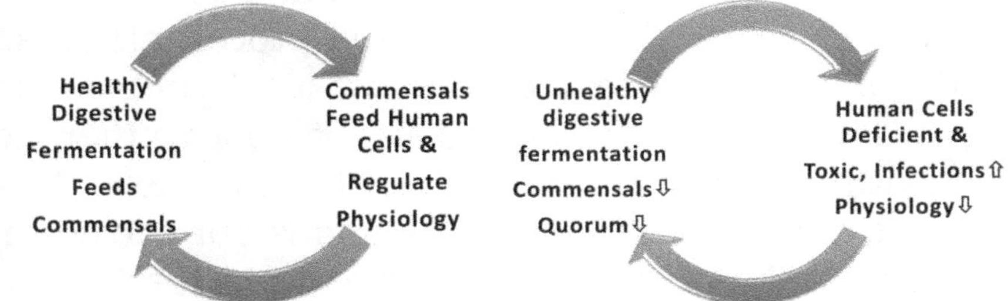

There is no question that when quorum communication breaks down, so does our health. Our bodies are continually under assault from a disruption of quorum communication. They telegraph the wrong messages causing immune cells to attack and turn on the body. This happens in fibromyalgia, auto-immune disease, cancer, Alzheimer's and nearly all diseases.

Today, nearly all disease is caused by man-made chemicals and pharmaceuticals that disrupt beneficial pro-quorum cell-to-cell communication and escalate dysbiotic or negative quorum cycles of infection and systemic toxicity. In clinical practice, this is where we find stubborn yeast and fungal infections, as well as runaway viral infections.

In 1800's, before the discovery of pharmaceuticals and man-made chemicals, man was more connected with nature, and only 1 in 200 people was diagnosed with cancer. In 1970, 1 in 15 died of cancer and today almost 1 in 2 people gets cancer. Diabetes has increased 700 percent since 1959, and all other degenerative disease rates are skyrocketing.

The main theme of this book is that all disease states begin at the cellular level with quorum disruptions that lead to a cascade effect, which ultimately ends up as some disease condition. Our cell colonies are losing their innate ability to coordinate processes well enough to achieve a pro-quorum "total body" effect.

In other words, a disrupted quorum process of human and commensal-probiotic cells makes it impossible for cells to communicate and benefit each other, so we slowly starve for nutrients and become toxic. This cycle is inherent in all of nature and is our ecological niche that allows us to thrive instead of experiencing distress and negative health consequences.

Numerous studies show that the majority of us have health vulnerabilities, and millions are eating years off their lives by obesity. I bet you know several dozen people who in the last year or two have followed diets, gone to support groups, or exercised with little or no positive effects.

The media and greedy marketers have distorted our thinking. Like looking through a prism, distorted information causes us to misperceive the true reality of why and how the body loses its innate ability to heal itself. Hundreds of studies show that your thinking and mindset directly influence your choices in life and your health status.

Let me be clear that this is *not* a book about medical diagnosis and medical treatment. Rather, it is about how you can use your inner physician to the fullest to enhance all realms of your life. Indeed, in my quest to study, observe, understand, and explain human suffering and illness and save my own life, I've found numerous concepts that present a provocative and new scientific perspective on why and how we get sick.

Engaging the genius of your inner physician and its coalition with nature is your best way to have better control over stress and your life situation. I've learned (and so will you) many specifics on how to get fully connected with nature and activate the most powerful healing force known to man.

The brain is the supreme center for sending and receiving signals crucial to quorum communication that contributes to the wholeness and cohesiveness of the human body. Dr. Jeremy Nicholson, a world renowned leader in commensal cell physiology at Imperial College London, feels that attention-deficit hyperactivity disorder, Tourette's syndrome, and autism all involve commensal cell deficits. "*We have some evidence now that shows that if you mess around with the gut microbes, you mess around with brain chemistry in major ways*," Nicholson remarks.

Getting all the cells in your body into an empowering environment is the foundation for enhancing your health and life according to nature. A recurring and important concept in understanding your inner physician and its marriage with nature is the quality of your environment.

Man-made chemicals and pharmaceuticals are synthetic and disrupt the healing power of nature. When we get overburdened with these toxins, we lose quorum cellular teamwork, a process that is vital for our survival in the 21st century. To solve our escalating national health crisis, we must restore a quorum between all the cells in the body by reconnecting with the battery-like polarities of nature.

Pollution is increasing faster than we can eliminate it from our bodies, at the rate of ten percent each year. We don't keep our internal bodies clean from the never-ending pollutant exposures. We put off exercising, overwork ourselves, and neglect to get enough rest and sleep.

The need to be nourished and healthy propels us to observe carefully what we eat and how we live our lives. Our satisfaction in life increases when we are able to live our lives in states of vibrant health.

The pursuit of good health requires looking inward. It means we need a new and deeper understanding of why we are ill or suffer needlessly from stress. Deep within our bodies at the quorum level is an awe-inspiring, prodigious, and miraculous system of cellular communication and innate healing and intelligence.

My consulting and clinical experiences over the past four decades have led me to the conclusion that most people are disconnected from nature's healing powers. Instead of being ruled by the genius of their inner physician, they stimulate and suppress symptoms with disastrous health results.

I've uncovered many of the reasons for this disconnection, but what is most shocking to me is the fact that people are completely clueless regarding their detachment from nature. Millions have derailed promising careers and lives because they are blind to the true source of healing within the body, the inner physician.

Like the wisdom that flows from nature itself, this healing entity can provide us with the ultimate energy and resource for sustaining both physical and mental well-being.

To tap into this energy, you need to consider every activity you do daily. Go through your day, beginning with when you wake, and take an inventory of your actions. As we go on this journey together, I implore you to reflect deeply on the following questions:

- ☐ Do I eat fresh raw foods that truly nourish my body?
- ☐ Do I find the time to supplement my diet with quorum nutritional concentrates derived solely from nature?
- ☐ Do I exercise, get adequate sunshine, and breathe deep during stressful situations?
- ☐ Do I drink good, clean, and harmoniously-balanced water?
- ☐ Do I think positive thoughts and align my emotions by expressing love and compassion to those I interact with?
- ☐ Do I compensate for fatigue and overwork with synthetic vitamins, sugar, or caffeine?

We are all engaged in the stress of life. Stress and human interaction can zap our vitality and drain our innate healing energies. The good news is that you can get in control of your life. You can be nourished, free of toxins, and stress-free. You can experience superhealing with proper nourishment.

When you engage your inner physician, you become your own source of healing; you gain your power and intuition back. You *stop* stimulating and *start* nourishing your innate healing mechanisms and your entire body.

Many who take supplements still feel sick and tired. Most people who believe they are taking nutritional supplements are actually consuming addictive, man-made chemicals or moldy powered supplements that over-stimulate the body. These supplements may even be labeled as "organic" and "natural."

Imagine wasting a ton of money on vitamins to trigger the very conditions you're trying to prevent! Imagine that these milligram-dosed, man-made supplements disconnect you with nature. Backed by over 35 years of scientific findings and my own research, *Quorum Nutrition*™ will teach you how to:

- ☐ Shed stubborn pounds in as little as one month by getting rid of fat-storing toxins**
- ☐ Enhance the cohesive function of all the cells in your body**
- ☐ Cleanse your body daily from harmful carcinogens to reduce your risk of cancer**
- ☐ Boost your energy levels without addictive caffeine or stimulants**
- ☐ Slow down—and even stop—a cycle of genetic illness in your family**
- ☐ Gain superior resilience to the nations number 1 killer...stress**

When I discovered how *Quorum Nutrition*™ works in harmony with nature and our inner physician, I was amazed. I wondered how something so simple and powerful is neglected by today's health care professionals. Later on, I made an even greater discovery regarding the "synergy" of quorum-fermented nutrients—working together—to activate extraordinary superhealing energies.

Did you know that the majority of today's supplements actually block innate healing and short-circuit your inner physician? This book presents the evidence and the argument for the potential of our inner physician and its innate intelligence to heal us. This amazing healing system is virtually untapped and malnourished in all of us. Yet, the power of this innate superhealing domain is gigantic.

Modern medicine doesn't teach you how to nourish, expand, and promote a greater activation of your inner physician, which can create super-resilience to today's number one killer: stress. The quorum process of nature holds the secret or key that unlocks the door for you to tap into your inner physician. Once this door is opened, you slowly break away from learned patterns of maladaptive behavior that diminish your innate healing abilities.

According to the World Health Organization (WHO), we rank the 37th worldwide in health care. In a recent 2007 DISCOVER magazine WHO stated: "*By any measure—longevity, infant mortality, burden of disease—we sit at the bottom of the basement of the industrialized world.*"

We face some daunting obstacles in activating the full potential of our innate healing capacities. Most of us will admit that we are not feeling truly vibrant, healthy, and physically fit. But for some odd reason, we don't listen and trust in our body's innate intelligence.

Could the fact that we're manipulated every day by the media and manufacturers of drugs and nutritional products be part of the problem? We are good at trusting people and believe what we are told and what we read. But, somehow in our life journey, we've lost trust in our body's own innate intelligence and self-healing capabilities.

What drives people, against their better judgment, to eat addictive and stimulatory foods? We starve, binge, and purge to battle excessive pounds. We devour diet advice, to little avail. We neglect the wisdom of our inner physician and its innate intelligence.

The food that flows through your digestive system can either empower you or wear you down physically and energetically. Food can maldigest or stagnate and become your primary source of stress and toxicity. Considering the fact that stress inhibits digestion, it becomes obvious why it is so hard to get nourished in today's stressful world. But in order for your digestive system to function smoothly and efficiently, you have to stop treating the symptoms and start addressing the root causes of illness.

You can change your food and supplement choices so that you nourish and strengthen your weak digestion and your inner physician's self-healing functions. But, you have to know what foods and supplements are truly connected with the healing power of nature. You have to acknowledge and understand that these malfunctions cannot be repaired with antacids, drugs, digestive enzymes, or herbs.

In a world that understands healing power as external, innate intelligence becomes short-circuited. Intuition is not regarded as important, and, therefore, is not processed. It is not expanded or made to function more efficiently from within the body at the cellular level. Just as there are techniques to enhance our intelligence with analytical thinking, studying, and repetition, there are techniques to engage and enhance our inner physician and source of innate intelligence or intuition.

- ☐ **The first of these is to honor, nurture and nourish your inner physician.** You must be willing to trust your body's inner physician to reveal and heal issues and then choose to work in harmony with it by eliminating toxic chemicals, addictions to sugar, coffee or alcohol and anything that can kill your commensal cells.

- ☐ **The second is use nourishment that restores quorum cell-to-cell communication.** As such, you will learn why it is far more important to implant lost commensal cells and then quickly nourish them with quorum-fermented nutrients. We cannot aim our nutritional efforts solely at human cells. You won't hear about this strategy from any of today's popular nutritionists, naturopaths, or doctors of alternative medicine. But the overwhelming truth is that commensal cell nourishment is a critical part of nature's plan for our inner physician to keep us vibrant and healthy.

- ☐ **The third step is cleanse your body of toxins on a daily basis.** Keeping your body's inner ecology healthy and cleansed will strengthen your inner physician and its connection with nature dramatically.

That's it! These three simple and powerful steps represent the most powerful nutrition I could ever share with you. That's the secret! And the key to opening the door to high level wellness is nourishment.

A chapter or more is devoted to each of these steps. Throughout this book, you will learn a lot about *why* following nature's wisdom in your dietary choices will help you to flatten your belly, fight fatigue and stress, and live a happy and active life.

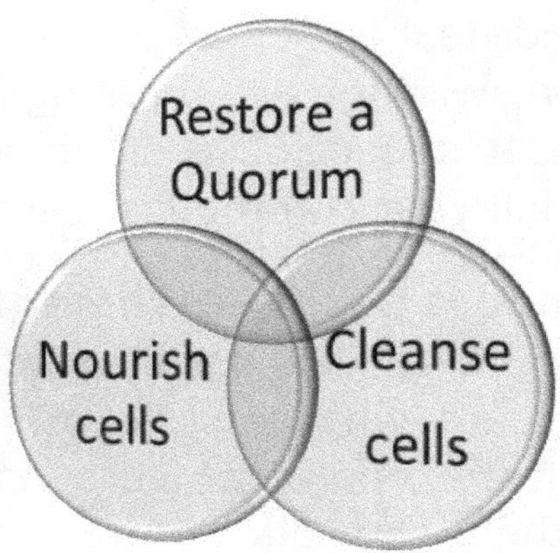

The last four decades of fad dieting have been linked to many chronic and life-threatening diseases. Here's a glimpse of the statistics:

- ☐ Over 18 million have diabetes, and millions are pre-diabetic and don't even know it.
- ☐ 50 million suffer from chronic allergic or inflammatory diseases.
- ☐ 70 million have arthritis and related disabilities.
- ☐ 80 million suffer from chronic digestive disorders, and millions do not even realize they are not digesting their foods and deriving nourishment from their food choices.

The air we breathe is so polluted that experts now say it kills three million people around the world every year. In the USA alone, it kills 100,000 people a year. Even more startling is the evidence that airborne pollutants increase the risk of a fatal heart attack or stroke (*New England Journal of Medicine, 2007; 356:447-58*). One study documented increased risk of death from respiratory causes (*Circulation, 2004,109:71-7*), while another showed a correlation with the degree of pollution and fatal heart attacks (*New England Journal of Medicine, 1993; 329: 1753-9*).

How can we protect ourselves from the health risks of breathing polluted air? *Quorum Nutrition*™ contains nature's synbiotic "pre-digested" nutrients to feed and energize cells into quorum states of miracle healing. Imagine getting raw, living nourishment; no pasteurization or nutrient-destroying chemical stabilizers or fillers, irradiated ingredients, flowing agents, binders, magnesium stearate, harmful mold or GMOs (commonly found in cultured or fermented whole-food products). These superior food concentrates can keep your body free of unwanted pollutants. A virtual goldmine of toxin-cleansing compounds can be found in fermented, synbiotic, food concentrates.

Remember, I believe your body has an inner physician or healer. It has the potential to emancipate you from needing to take pills to spike your energy levels or to take drugs to suppress your symptoms. It will free you from the way you have been living your life (in ignorance of your innate healing potential).

This book is about living a life that is filled with balance, energy, and stamina. Part of the reason I'm giving you this new information is so that you'll realize the importance of keeping your innate healing system in good working order.

If you want to prevent the diseases that cut life short, *Quorum Nutrition*™ is your ally in two ways. First, this type of nutrition quickly corrects deficiency states that have been linked to almost every type of degenerative disease, and it nourishes commensal cells so they can perform miraculous healing feats in your body.** Second, it comes with an unbelievable bonus: It cleanses the body of toxins, normalizes hormones, promotes innate healing, and makes you incredibly stress resilient.**

Simply put, my discovery of nature's quorum healing secrets can release powerful streams of healing energy within you. It's living from the inside out—finding your inner physician and letting it work optimally in a way that can leverage your suppressed healing abilities and innate intelligence. When you have an illness, your inner physician remains short-circuited and untapped, buried somewhere deep inside. This is an awful way to live. And, it drags down those you love and care about—your family and friends.

Life shouldn't be that way—and it doesn't have to be. When you release your body's potential, you'll feel alive, connected, and full of vitality. Your family, friends, and co-workers will all benefit from the positive transformation in your life.

So stop playing the victim, and summon the courage within you to get on with your life. Overcoming fear and hopelessness and connecting more fully with nature is crucial for you to succeed on this journey and will unlock the vault where your body's inner physician is hidden.

You're about to embark on an educational journey that will teach you how to de-stress and enhance your body's innate wisdom and healing capacity. You'll take charge of your health and be in control of stress. But, remember: Many of your ailments are due to years of failing to eat wisely, exercise appropriately, or nourish your body in ways that enhance your body's innate healing energies.

You cannot augment your innate healer unless you detach from addictive habits that cause a yo-yo effect with your emotions, blood sugar, and energy levels. Each symptom of excessive or low energy is a message for you, a signal from inner physician about a critical need nourishment and not stimulation. This means you have to learn how to STOP suppressing these innate messages with synthetic nutrition or drugs and START answering them with *Quorum Nutrition™.* When you look as these symptoms as obstacles or bodily expressions that need to be controlled, you miss the point. When you ignore them, you ignore your inner physician and its genius intelligence.

Having survived two terminal illnesses, I discovered the combination to the vault where my inner physician was kept locked from a childhood of prescription medications. Discovering these precious innate healing gifts within saved my own life twice and unlocked the vault so I could gain full access to my inner physician. My mission is to teach you what I've learned and applied successfully on myself and thousands of patients in conjunction with medical physicians so you may reawaken your inner physician which has been damaged due to neglect and environmental toxicity. Only you can unlock the vault, as it must be opened from the inside out.

You need to look at your inner physician as your best friend, a friend that stays with your through the most difficult times. Your innate intelligence will never leave you, but the more you neglect to trust and listen to it, the more you will subdue your inner healer. Above all, recognizing that your health issues come from malnourishment and toxicity is important.

Trusting the awesome intelligence of your body means you respect its gifts of imparted wisdom by providing your body with ample nourishment. You savor those gifts and hold them precious to your well-being. Without this intelligence, you cannot improve your health and well-being.

Your inner physician uses nature's handiwork in inexplicably producing spontaneous remissions in many illnesses. It's incredible innate intelligence works through the body's energetic and neurological circuitry. Indeed, genuine cases exist, and throughout the history of medicine physicians have recorded cases of spontaneous remission in cancer cases and all sorts of diseases.

Do not allow yourself to be influenced by someone else's ideas or narrow-minded opinions. Most importantly, do not allow people or organizations to take your money for dead-end research that never comes up with real solutions. False hope is a trap designed to keep you enslaved to the empty philosophies of greedy people. Unfortunately, many people who read this book will never follow through with all of the eight recommended steps. As a result, they will continue to plod along through the same hopeless, depressed, and sub-optimal life.

You have to take care of yourself. Your body's inner physician runs your life. When it is out of tune, it deteriorates, and nerves that control digestion and immune responses become worn out and short-circuited. The function of your inner physician is as essential for our well-being and growth as sunshine, clean air, and water. While making changes in your life is never easy, focusing on the bigger picture and other rewards you get from this program will keep you moving forward. Sure, there may be setbacks, but the health benefits will far outweigh them, and pretty soon you will be amazed how healthy your life has become.

If you are not eating the right foods, drinking good quality water, exercising consistently, and getting plenty of nourishment and rest, your body's inner physician is breaking down. If wellness is not at the forefront of your life, I can almost guarantee you that your health issues will end up causing cancer, a heart attack or stroke, or premature death.

As you will learn. when nourishment is poor and stress is high for many years, it takes a unique kind of nourishment to repair our damaged digestive tract and reawaken the full capacity of innate healing. It must be in a *quorum* format. Quorum-fermented nutrients are created by nature through the miracle of fermentation. These nutrients must activate and enliven commensals, your body's miracle healing cells. You'll learn more about this unique kind of nutrition called *Quorum Nutrition*™ later in this book, and you'll benefit from new ways to experience a higher level of health and wellness.

When writing this book, I have always kept you, the reader, in mind. I want nothing less than your health and happiness. But you have to be a partner with me. You have to take the recommendations seriously and take part in your own improved health regimen. This probably means you must change—change the way you shop, the way you eat, the way you exercise, and even the way you think. But I am convinced it will be worth it, because it will be a vast change for the better!

Keep reading, learning, and questioning the health myths you face every day. Live your life in superhealth and not enslaved to dangerous pharmaceuticals or megavitamins or symptom-chasing therapies. Allow your body's inner physician guided healing ability to double, triple, and expand its boundaries to make extraordinary health and wellness a reality in your life so you can feel a vitality you have never felt before.

Dr. Paul Yanick Jr

Chapter One

The Genius of Nature and its Quorum Cycles

Imagine having the power to change the world and transform your own health. As simple as it sounds, the tips and ideas you'll read about have a lot behind them. Each assertion made about the potential hazards of man-made chemicals, drugs, nutritional and herbal products, and nutritional practices is my opinion, and is based on prestigious, scientific studies and the hard lessons I learned in saving my own life twice. Behind these assertions lies over four decades of painstaking research and a deep base of knowledge and respect for natural laws and living an eco-friendly life that causes no permanent damage to the planet or the human body.

Our inner physician is one of the most energy-hungry systems of the body. It is not hungry for calories or man-made nutrition—It goes much deeper than that. The hunger is for life-sustaining nourishment at the cell level. And, it is nourishment in the right format that helps the inner physician function optimally. That is the dominant theme of this entire e-book.

Today, many are scared. And with good reason—there is too much cancer, too much obesity, too much heart disease, and too many living into old age without quality of life. Because most of us lack knowledge of how to nourish our bodies, food and diet marketers are having a field day with our ignorance. Adding to this confusion, science and medicine contradicts itself every other month.

A recent associated press release stated that 34 people were hospitalized and 110 people were medically treated after a co-worker spayed perfume. Every day we are exposed to the toxic fragrances that people wear on their bodies or use in cleaning their clothes or home. To protect ourselves, we have to restore our body's self-cleaning ecosystem.

Reclaim a Quorum of Nature in your Life

A quorum of all the specialized cells in your body means there is cooperation and efficient function of your cellular anatomy. Most importantly, achieving a quorum gives you nurturing and protection you need from a toxic world.

One of the best ways to understand nature's quorum secrets is to look at the amazing cooperation and teamwork among insects that stay tightly connected to the healing powers of nature. Colonies of the ant *Temnothorax albipennis* nest in small crevices between rocks, and when the rocks shift and the nest is broken or overcrowded, they use quorum communication in the process of choosing an optimum new nesting site (*Proceedings of the Royal Society B-Biological*

Sciences 273, 2006; Biology Letters 1:2,2005; Behavioral Ecology and Sociobiology 50:4: 2001). In a coordinated and highly intelligent fashion the new recruits visit potential nest sites, making assessments of their quality. The number of ants visiting the nest increases until a quorum has been met, and the entire colony has been restored (*Behavioral Ecology 16 :2, 2005*).

It is a quorum status at the nest site that signals carrying of the brood, queen, and fellow workers to the new nest. There is a strong support system and amazing teamwork for the mutual benefit of all the ants in the colony.

Honey bees (*Apis mellifera*) also use quorum sensing to make decisions about new nest sites by forming swarms that hang from a branch or overhanging. This swarm persists during the quorum decision-making phase until a new nest site is chosen (*Apidologie 35:2, 2004; American Scientist 94:3, 2006*).

The body's inner physician resides in the brain and neurons. Scientists at the University of Bristol study ant colonies because ants and neurons exist simultaneously as an individual and a collective, quorum working team. With a grant from the UK's Biotechnology and Biological Sciences Research Council, Dr. James Marshall and biologist Nigel Franks are noting systemic similarities and physical incongruities of these two systems to illuminate general principles of quorum group decision-making, a process that pervades biology and society at all levels of life.

Through this research, Dr. Marshall says he hopes to "*elucidate general principles of decision-making by comparing the processes and structures of social insects with those in vertebrate brains, such as in the primate visual cortex*." A decision is made when a "quorum" is reached, when a certain number of ants agree on a location, and the same process occurs among neurons in a monkey's visual cortex when the animal performs a visual discrimination task. In the task, a monkey is flashed an image of dots moving in different directions and must decide which way the majority of dots are moving. Amazingly, when the image appears, quorum sensing in the neurons of the monkey's visual cortex gather bits of information from the monkey's eyes, much like ants evaluating a nest site. As more data is gathered, the neurons with the correct answer gradually increase their firing rate and an intelligent decsion is made .

Neurons in humans can adjust for either speed or accuracy when a quorum of all body cells is met. The efficiency of a social insect colony or a neuron-based network of nerves is largely a product of the simplicity of its constituents and its inherent connection to nature. Humans have conflicting agendas, whereas social insects in a colony have much less self-interest and genuinely want to converge on the best option for the entire social network.

The similarities between ants and neurons suggest there are general principles of quorum teamwork among cells that can make us far smarter and far more efficient. By looking at our quorum connection to nature, we have to step outside of standard modes of reasoning and health care and find new ways to talk about the powerful synergy between our ecosystems and our cells.

Thus, self organization of our cells against invading microbes or pollutants and inner physician healing is mediated by quorum sensing. It is this amazing process that allows the body's inner physician to empower itself to keep us at peak levels of health no matter what is stressing us out.

Barometers for a Disruption of Nature in our Lives

Just like chronic fatigue tells you it's time to rest and get needed sleep, your inner physican wants you to confront what it is that keeps you stuck in an unhealthy environment that fosters a disconnection from nature. Only you know how you truly feel and if you are honestly answering questions such as these: Do I feel good and energetic without taking any stimulants like caffeine? Am I experiencing reduced physical health? Do I struggle with anxiety or nervousness and binge eating?

Barometers for assessing your connection with nature are empowering to your inner physician. The more intensely and frequently you connect with them, the more vibrant your health will be, the truer the match.

As I said, the interaction between your inner physician and nature is the key to health. The more favorable the match, the more powerful your inner physician functions. Putting yourself into a nurturing and nourishing environment will help you develop and establish quourm cell-to-cell communication that keeps all the systems of your body in sync.

Be tuned to the fact that we often hide from our true biological needs. When we fail to acknowledge what truly makes us feel good and energetic, we foster a disconnection with nature. Our body functions in an ill-suited environment that cannot keep our cells in quorum states. The habitat that continulally promotes this connection with nature must be approached at commensal cell physiology.

Think of yourself as a walking processing system. Air flows into your lungs and then flows out of you. Food-based nutrients, water, and toxins flow in and out of you. And, the level of energy you process from these interactions can be positive or negative in terms of innate healing.

Currents of energy run through you via energy circuits called acupuncture meridians and neurons. Life is governed by this neuron energy flow, which also governs our physical functions (physiology). But to heal properly, energy must flow efficiently with the correct polarity, and neurons need quorum nourishment from commensal cells. Your life depends on nourishment and the transmission of neural energy.

On every level of our being, science has documented that we all have valid capacities of self-healing. This supreme force in our physical nature can provide us with the ultimate energy or stab us with endless pain and inflammation. The fact is that in today's modern world, our energy systems commonly fail to function smoothly and effortlessly. As you will learn, the primary reason for this malfunction is that we have separated ourselves from the healing power of nature.

A lack of nature in our lives causes nerve energy to become turbulent, blocked, and inefficient at keeping our bodies healthy. When energy is turbulent, we become angry, fearful, and may even gain weight uncontrollably. This causes many to focus on negative emotions.

It is well established that STRESS and our ENVIRONMENT (malnourishment or toxicity) are responsible for almost all of today's illnesses. To reap the benefits of this program, you don't have to engage in excessive exercise or stress reduction techniques. As I will explain further in this book, it is *quorum nutrition*™ that will tame stress and stop it from damaging your body.

Despite spending billions of dollars on diet fads and nutritional supplements, the majority of people are not healthy and not at a healthy weight. They skimp, fast, avoid fat or carbohydrates, and drink endless meal replacement beverages. Yet, despite their efforts, they manage to gain back every pound they lost. Now imagine a weight loss program designed to change all that. By embracing nature's plan for health, it is easy to lose weight and get healthy and stress-resilient. It's the opposite of the quick-fix approach—and it really works!

Today's obesity epidemic is caused by a nation of people who are addicted to foods and stimulation. We make dietary choices that make us fat, generate inflammation, and kill us before our time. Of all the things I have learned in the past 35 years, nothing is more important to promote innate healing than how we get our nourishment. We can choose to eat ourselves into a state of obesity, fatigue, and inflammation (pain) or to experience bountiful energy and vibrant health by connecting more and more with nature.

When we think about the healing power of nature that flows through us, it is prodigious, awe-inspiring, and miraculous. For example, when a salamander's leg is torn off by a predator, it simply grows a new one. When a pond worm is chopped in half, each half regrows into a complete worm. How do these organisms know how and where to reassemble new organs?

The miraculous intelligence and power of regrowth and regeneration are inherent in their anatomy and connection with nature.

Scientists have found this kind of healing power in humans when we are embryos. The force and magnitude and the proper regulation of this flow of intelligent healing energy determines the level of health when we are born. But, as countless studies document, we are born toxic. We acquire the toxicities of our parents, and we quickly become enslaved to a dependence on drugs rather than nature. Thereafter, our bodies seem to engage in prolonged warfare against all sort of pathogens, toxic chemicals, and other stressors.

It is our environment that disconnects us from nature and overengages and overextends our genes. Toxins can shut genes off or mutate them, causing them to make *misfolded proteins*. When genes are disrupted, a cell can turn cancerous or die faster than the body can renew it. Since many of us today are born toxic and nutrient-depleted, our healing genes tend to dysfunction. We don't always rebuild tissue or keep our cells in equilbrium fast enough to maintain our bodies at a peak physical state.

Is Your Body A Garbage Dump?

Consider the consequences of a garbage strike in a city. Garbage accumulates, streets are clogged, disease spreads, and daily life is disputed. Eventually, things can come to a standstill.

Dr. Fred Cohen of University of California, San Francisco, says it is like a garbage strike, "*...the trash on the sidewalk begins to stink. That's what we're* *dealing with here."* Scientists like Dr. Peter Lansbury at Harvard Medical School feel that misfolded proteins lie at the heart of a wide array of diseases afflicting millions of Americans. And, more and more research is pointing to nutrient deficiencies and environmental toxins as the main cause of these genetic aberrations. Dr. Cohen says, *"Millions of people might escape the ravages of these really insidious diseases,"* when we understand how the body's trash removal system fails.

Believe it or not, despite these genetic shortcomings, our capability for inner healing is so great that to explore the endless boundaries of its power would render helpless even the best of minds. The shortcoming of scientific study of the human body is that it neglects to study this hidden dimension. Instead, scientists search for answers in our chemistry and molecular makeup.

Very few scientists are actually asking the question: How can the inner physician be nourished, expanded, or function more efficiently? The answers to this question are starting to come together in a variety of scientific disiplines. By following the path of our embryologic origin, scientists are mapping out healing networks that connect to our DNA and the neuron that serve as the prime tool used by our inner physician.

Keeping the Life Principle in our Daily Life

Biochemistry alone cannot explain life. Leading biochemists have documented that chemical reactions take place billions of times faster in the living body than in a test tube. This one fact should be sufficient to spur recognition of the critical importance of our innate healing intelligence—what we could rightly call the "*life principle*."

Biochemists have encountered innumerable failures when they've tried to synthesize vitamins or the building blocks of life or create life in a test tube. Without taking the life principle into consideration, they are doomed to even more failures.

Both buffeted and compelled by the rewards of political, medical, and pharmaceutical industry economics, many researchers neglect the natural healing dimension of the body. This has been an unfortunate deterrent to understanding how to nurture, strengthen, and fulfill the body's healing abilities. But, while researchers and mainstream doctors have their heads buried in the sand, ordinary people everywhere have begun to look inward to discover the long-hidden powers of the body. There is an ever-widening interest in alternative medicine and natural healing methods.

As the founder of *Quantum Medicine*™ in 1981, I've made a solid case for the existence of a superior intelligence with unimagined potency and range of healing powers within every person. From the sum of evidence, I believe, emerges a quorum superhealing potential. To achieve this potential, we have to respect and connect to nature's wisdom.

By incorporating nature's plan, we can meet all health challenges. Nature's plan works in many forms of life. Consider ants that have more collective biomass on Earth that humans do, and inhabit as diverse a range of environments. Ants are able to hunt, grow, and ferment their own food; build their own homes and underground cities; effectively handle their wastes; create powerful medicine; and produce biological and chemical weapons without polluting the earth. Consider the swarm intelligence of the honey bee. But, one has to wonder if environmental pollution was the cause of a massive death of honey bees in recent years? Air, water, and food pollution are taking a toll on our life as they disrupt our quorum cycles and nourishment. There is too much CO^2 and not enough oxygen and nutrients.

Since nourishment is the key to restoring the body's inner physician, we have to carefully examine the kind of nourishment we use in everyday life. In 1981 I founded *Quantum Nutrition*® and *Quantum Medicine*™ as a way to teach and board certify clinicians so they could connect their patients with the healing power of nature. With Quantum Nutrition, too many jumped on the bandwagon of my earlier discoveries, putting their own twist and ideas on my original ideas. This created chaos and confusion as many clinicians were representing that they were doing my methodologies.

Now, with the discovery of *Quorum Nutrition*™, people will find the original and authentic source of nutrition that I researched to save my own life and the lives of thousands who sought the help of physicians whom I trained.

Even the distinguished physician and Nobel laureate, Albert Von Szent-Gyorgyi, MD, respected the nourishment from nature, stating, "*The deeper we go into the facets of life, the more mysteries we encounter. Analyzing living systems, we often have to pull them into pieces, decompose complex biological happenings into simple reactions. The smaller and simpler the system we study, the more it will satisfy the rules of physics and chemistry, the more we will understand it, but also the less 'alive' it will be.*

So when we have broken down living systems into molecules and analyzed their behavior, we may kid ourselves into believing that we know what life is, forgetting that molecules have no life at all."

Living cells need living nourishment. Through the miracle of quorum fermentation, nature choreographs some wonderfully elegant and extremely complex synbiotic nourishment that makes man-made efforts in the laboratory look incomplete and awkwardly designed. This form of raw, living nourishment delivers thousands of living cell nutrients and has the added bonus of a long shelf life. Think about it. Manufactures of powered nutritional products or powdered herbs and superfoods want you to believe these products have a long shelf life and last for months and even years. Researchers have monitored and watched these dried foods and antioxidant juices deteriorate in nutrient content and increase in mold just weeks and months after being bottled.

Vitamin D is called the "sunshine vitamin" because it's not only found in food, but can be made in your body after exposure to ultraviolet rays from the sun. The sun and the miracle of photosynthesis, along with soil bacteria, provide the ultimate in organic minerals and vitamin D. If people spent more time in the sun and stopped using excessive sun blocks, there would be no such thing as vitamin D and calcium deficiencies (vitamin D is needed to get calcium into your bones). When the dairy industry started pasteurizing dairy products, they made it impossible for us to get calcium from milk. And, now they wonder why America has a bone health crisis!

Before you reach for your sun block, realize that the sun can only damage nutrient-deficient skin.

Quorum nutrition™ can keep your skin strong and healthy. Aborigines in Australia who live outdoors don't get melanoma and die because they do not use sunscreen. They are closely linked to nature's quorum symbiotic cycles and get the kind of nourishment that keeps their skin healthy.

The way we think about our health is costing us money and our lives. The health care model is predicated on false beliefs, misinformation, outdated ideas, greed, and fear. Doctors want us to believe that our self-healing resources are scarce and that we have to control the body with man-made chemicals like pharmaceuticals to bring it back to health.

In the thrall of this scarcity mindset, we make faulty health decisions. We buy and consume products that are destroying our inner physician. The greed of big business blinds us. This mindset is characterized by an adversarial relationship with our inner physician and nature. Even nature is suffering by this mindset as scientists are in agreement on global warming and the negative consequences of pollution on the earth.

If you are sick, you must see a licensed physician who does not use man-made pharmaceuticals and nutraceuticals and who respects and wants to work with your inner physician. In the next chapter you will learn more about how our ecosystems are threatened and extremely challenged in today's day and age.

Chapter Two

Your Personal Quorum Evaluation

Although myths regarding health create dilemmas, these dilemmas are easy to overcome once we educate ourselves to the indisputable truths found in nature. In this chapter, you'll explore your own health status. With the right knowledge, approach, process, and tools, you will discover how to unlock the brilliance of your inner physician.

As Norman Cousins stated, "*The marvelous pharmacy that was designed by nature and placed into our being by the universal architect produces most of the medicines we need*." Since stress is a thief of health, using nature's pharmacy of commensal cell generated quorum nutrition can help immensely to protect us against stress and our toxic environment.

Each toxin we breathe, and we breathe hundreds of them daily, is a stressor to our inner physician. All these stressors are elusive, stealthy, capricious, pernicious, deceptive, and dangerous.

What about everyday life stress? Stress, and our reaction to it, can overwork the body until our inner physician becomes totally depleted. It is everywhere, intertwined with the rhythm of life. Coping with kids, budgets and jobs, gardens and cars, shopping and schools, bills and repairs are everyday sources of stress.

Even small stresses cause a pile-up of tensions that constricts the flow of healing energy in our bodies. Too much or too frequent stresses add up to zap our healing energies, allowing stress to damage the body.

When you think of stress, you have to realize that there are chemical, electromagnetic, and even dietary stressors that impact us daily. Chemical and electromagnetic pollution have doubled just in the past 3 years. And, stress depletes our cellular nutrients, the vital substances of life and diminishes circulation and nervous system regulation of digestion and detoxification. If we are stressed out, it means that no matter how good or how much we eat, we cannot adequately nourish the cells in our bodies.

Man-made chemicals have complicated our lives and elevated an already dangerous level of stress present with today's lifestyle. Like a terrorist lurking in the shadows, these toxins lurk in our bodies, hiding like a ticking time bomb, eroding our vitality and waiting for our weakest moment to attack; a devastation which many will never survive. Awareness and understanding of how our inner physician and living ecosystem weakens and damages our stress defense mechanisms can prevent stress from wrenching the emotions, deranging the mind, impairing the body, and extinguishing life. The first step in understanding how stress is affecting your body requires a self analysis. I've created a list of symptoms that will aid you in evaluating your own level of stress.

Your answers will give you some insight into how much you are suppressing your innate healing capacities. Do you... (*circle each statement that applies to you*)

- ☐ Crave starches and sweets?
- ☐ Have problems losing weight and have excess belly fat?
- ☐ Have low sexual energy?
- ☐ Have difficulty remembering things?
- ☐ Have periods of anxiety?
- ☐ Frequently get constipated?
- ☐ Fail to eat a good, healthy breakfast?
- ☐ Eat white bread or pasta daily?
- ☐ Chew gum on a regular basis?
- ☐ Drink juices, soda, or other sweet drinks?
- ☐ Need a cup of coffee or tea to get going every morning?
- ☐ Eat margarine in place of butter?
- ☐ Tend to eat local meals and drink diet sodas?
- ☐ Eat fried foods daily?
- ☐ Binge on sweets more than once a week?
- ☐ Feel tired late in the afternoon?
- ☐ Feel sleepy after dinner?
- ☐ Find it's hard to stay focused at work?

☐ Frequently get headaches?

☐ Frequently feel lightheaded?

☐ Yawn a lot during the day?

☐ Feel depressed or sad once a week or more?

☐ Have allergies?

☐ Wake up tired in the morning?

☐ Feel nervous and irritable?

☐ Drink any alcoholic beverages daily?

☐ Have trouble falling asleep?

☐ Wake up at three or four in the AM and can't sleep?

☐ Experience muscle pain or spasms?

☐ Have low or high blood pressure?

If you answered "yes" to more than six of these questions, you could very well be experiencing a 40% quorum deficit in the functional status of your inner physician, and you have a mild to moderate dysbiosis, toxicity, and malnourishment at the cell level.

If you've answered "yes" to more than twelve of these questions, your inner physician is under chronic stress. Your ecosystem is likely experiencing moderate to severe dysbiosis and uncompensated stress and stored toxins are damaging your body. Your body can't detoxify pollutants effectively, your stress-fighting hormones are out of balance, and your efferent neurons are pushing your immune system into overdrive. Inflammation is excessive and chronic. Your nervous system is depleted and toxic and not able to keep your body balanced. Commensal cells are woefully deficient, and your body is in a perpetual distress cycle.

Here's the good news: I've discovered ways that can help you break free of these vicious cycles. Changing dietary habits that impede your health can alleviate your discomforts and limitations and make you happier and more productive.

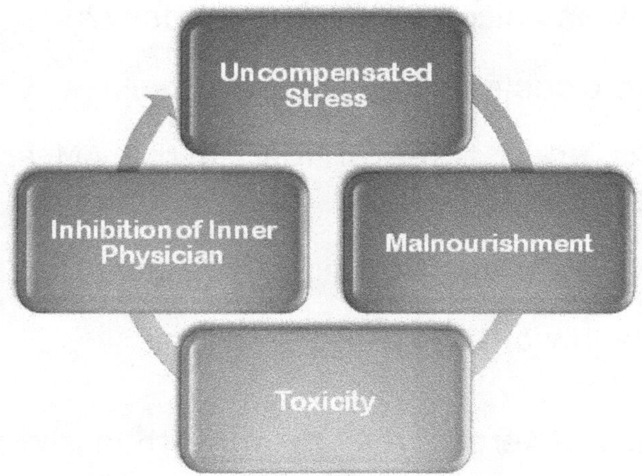

You may not realize that uncompensated stress inhibits your inner physician or that the best way to conquer stress is with quorum nutrition™. If you're not adequately nourished and/or are too toxic you are constantly draining the power of your innate healing mechanisms. This means you have little or no stress-protective defenses in operation. Stress becomes distress and damages your body.

Prolonged and uncompensated stress causes elevations in the stress hormones called cortisone and cortisol and weakens the afferent neurons that promote regeneration and increase nourishment. On the opposite end of the see-saw of efferent-afferent neuron balance is the efferent neurons. As they get over engaged, we lost digestive and detoxification functions and turn on the fires of inflammation.

These hormones inhibit cells that act like stem cells called *fibroblasts*. When fibroblast cell activity is diminished, our bodies stiffen. The motion between the muscles, blood and lymph vessels, organs and nerves gets out of sync. The stiffer our connective tissue gets, the less we heal. There is a constant deferral of innate healing and repair routines in the body, so stress-induced damage cannot be repaired fast enough by the body. We age prematurely, can't lose weight, and feel tired all the time.

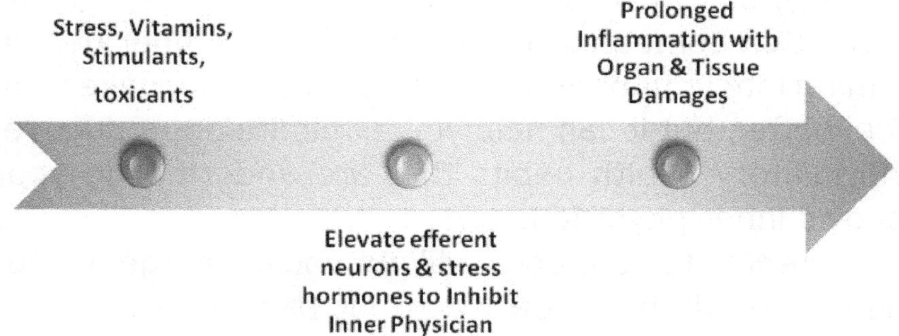

Reducing the stress of life is not simply a matter of dissolving tensions with relaxation techniques. Stress illnesses are the result of progressive malnourishment and toxicity that weaken the body's stress defense mechanisms.

Stress is not solely emotional in origin. Most of the stress in our lives involves nutritional deficiencies that leave us feeling physically, emotionally, and mentally overwhelmed and overloaded. Everyday life routines like juggling family responsibilities while trying to satisfy the demands of a career are not stressful when nutrition is optimal and stress is compensated for.

I want you to feel like a winner on this program, but remember one thing. You must trust and have deep respect for your body's inner physician. Once you fully trust in your body's ability to heal itself, you will have boundless energy and an improved sense of well-being.

The secret of stress resilience, an ability to roll with the punches, is to fully compensate for stress with the right nourishment. Nature has a way of guiding us toward making right choices when we stop stimulating and start nourishing our body's healing domain.

Such symptoms as pain, swelling, aching, nausea, anxiety, skin rashes, a stuffy nose, or fatigue are all signs that something is wrong with your inner healer. Tuning into your body and its needs not only makes you feel better, but it can help you eliminate unproductive, stimulatory health habits that are undermining your body's inner physician.

Listening to your body will put you in charge of your health. It will help you weed out bad dietary habits. It will help to break habits that weaken your inner physician.

Self awareness is the beginning of wisdom and is a prerequisite for self-healing. In the hustle and bustle of everyday life, we tend to forget to listen to our body language. Instead, we take a pill to calm our stomach or to provide headache or pain relief. So many people don't start listening until something happens that jolts them out of their lazy, addictive, unhealthy habits. I encourage you to listen before it is too late to fix your health problem. Remember, fifty percent of people who have a heart attack don't survive, and cancer is now killing one out of three Americans.

With the information I've compiled, you can control the optimum function of inner physician controlled afferent neurons. You can even improve your body's production of anti-aging hormones: hormones that put the brakes on allergic reactions, boost immunity, stop pain and inflammation, improve digestion, and stimulate the manufacture and secretion of vital hormones to give you endless energy.** You can minimize the release of stress-generated hormones that cause accelerated aging in your body.**

Since stress damage originates from poorly nourished cells, having a form of nourishment that provides an instant supply of stress-defending nutrients to our commensal cells and neurons is critical in awakening healing and preventing the harmful effects of stress.

Stress is produced by any stressor that robs the body of wholeness. Remember, stress diminishes wholeness at the core-link of our embryologic origin, causing our inner physician to get weak and dysfunctional.

Today's propaganda requires people to stop thinking, to accept social norms and clichés unquestionably, and to quickly consent to all sorts of health damaging practices. This blind trust of medicine that is prompted by fear and misinformation is a form of gambling and not investing in our health. It blinds us from seeing the unlimited potential of our inner physician.

If we want to maximize our health potential and live our life full of happiness and vitality, we must see through the deceptions of the commercial health and nutritional industry. Focusing on treating symptoms is like harvesting the fruits of a tree while ignoring its roots. We have to trace the fruits back to the branches, then to the roots to see what made the fruits grow in the first place.

If we want our health to prosper, we must focus on reducing or eliminating toxic man-made chemicals that kill our commensal cells and diminish the superhealing powers of our inner physician. Once you have identified root causes for inadequate inner healing, the next step is to learn the lessons and never repeat the mistakes. I've learned this all the hard way myself. For me to survive two life-threatening illnesses, I had to concern myself with the root causes of my disease, rather than hacking at the byproducts or symptoms of my disease. I'll share more of this story with you later on in the next chapter.

I believe that too many are trapped inside the destructive power of health myths and stuck in degenerative anti-nature cycles (dysbiosis). Deep down inside we know something more is wrong than what we're being told.

Are You Overweight?

Obesity is rapidly becoming a national epidemic and is one major indication that our inner physician is distressed. Americans are getting fatter—a disturbing trend because obesity equates to an increased risk for diabetes, cardiovascular disease, cancer, and other chronic diseases.

Maintaining a healthy and safe weight begins by not following one diet fad after another, but by understanding the depleted nutriture underlying obesity. While getting adequate nourishment is the logical place to start, we must realize that, by the time an individual is obese, he or she cannot digest food efficiently and there is dysbiosis. With a loss of quorum communication between our human and commensal cells, we fail to produce and activate the fat-burning thyroid hormone, T_3.

Fad diets come and go. Low-carb (glycemic index), low-fat, liquid diets, the grapefruit diet, fat-flush plans, and starvation diets all have limited success rates in the long-term. People take digestive enzymes, spike energy with USP vitamins, herbal stimulants, or caffeine-containing beverages. All these practices only serve to weaken digestion and lower cellular nutrient uptake to intensify starvation signals to the brain. When this happens, the body cannot get enough nourishment to fuel metabolism and quell the disease-causing fires of inflammation, and nutrient uptake is further diminished.

Are there secrets to avoiding obesity? Healthy eating requires consideration of much more than diet and exercise. If you're looking to lose weight and build a strong, lean body, you have to nourish your body's commensal cells so they can get rid of fat-storing toxins (xenoestrogens) that block fat-burning hormones. Later on you will learn more about toxins called *xenoestrogens,* the root cause of obesity.

Quorum Nutrition™ can jump-start your metabolism to ignite powerful fat-burning metabolic pathways. Later on, I will also show how a lack of exercise and poor posture can result in a larger and weaker abdomen. This information will include simple posture changes that you can use while fast walking to re-align the body and improve the operational complexity of your inner physician. Combining simple alignment exercises with brisk walking burns about 300 calories in only 35 minutes compared to walking alone, which burns calories at the rate of 270 calories per hour. .

Widespread obesity and malnourishment exist side by side. The world had become a place simultaneously of overabundance and starvation. With obesity rates soaring, it's important to understand what kind of nourishment is best for the body. Taking a daily multivitamin supplement can have a profound effect, for better or worse, on how your body functions.

Are you Immune Compromised?

In the war against cancer and other stress-associated disorders, scientists have identified certain foods that are among the richest sources of active nutrients. These sources provide nourishment that is hundreds, or maybe even thousands, of times greater than conventional foods and far surpasses common nutritional supplements, especially when fermented by pro-quorum tactics.

Keeping your immune system strong is paramount for good health and longevity. Letting your immune defenses fail allows you to become a target for a host of health problems, including cancer. Quorum Nutrition™ works by creating a potent jump start for your internal commensal cell and immune cell army. When properly nourished, these cells go into immediate warfare with hidden bacteria and viruses in your body.

Anyone with a family history or a high risk of cancer, or who's worried about getting cancer, should consider taking quorum nutrition™ in place of their daily multivitamin-mineral supplement to nourish and support their immune system.**

Do You Have Hormonal Imbalances?

Quorum nutrients are far superior to synthetic or man-made vitamins or soil-derived minerals. Why? Because they are biologically active and easier to digest and assimilate into our cells. And, most importanly, because they tend to keep our body's hormones on an even keel when we're under stress. When stress hormones are restrained, fibroblast cells that act like stem cells to regenerate the body, are not inhibited.**

Remember that fibroblast cells are responsible for the free-flowing motion between organs, muscles, vessels, and nerve-based systems of the body. Like telephone lines that carry signals around the world in seconds, your body's connective tissues carry the signals of your inner physician throughout the body. In this manner, your neurons can receive information clearly, respond to it quickly, and be fully operational in healing the body. The wholeness or functional unity of the body is restored. This means that all the systems of the body are in sync. Quorum nutrients are stress protector nutrients that have the amazing poteintal to sharpen the mind and boost the body's metabolism, resulting in more muscle and less fat.** These powerful nutrients promise to extend the human life span and give us a chance to live a full, productive, and more youthful life. Imagine being able to maintain the peak hormone levels or immunity of a twenty-year-old as you grow older.

Alice's Struggle with Irritable Bowel Syndrome

Alice is one of millions of Americans who was diagnosed with Irritable Bowel Syndrome (IBS), a condition always lacking commensal cells and quorum nutrients. IBS is associated with dysfunctional patterns in bowel motility, how the smooth muscle in the wall of the intestine contracts and relaxes to move the digesting food through the intestines. Many people don't know that IBS is associated with a lack of beneficial commensal cells in the body. If, like Alice, you've been hit with symptoms of fatigue or chronic digestive disorders, and medical doctors can't diagnose your problem, start thinking about the possibility that you need *Quorum Nutrition*™.

Most Americans don't get all the nutrients they need from their overcooked, over-salted, and over-processed food. Alice's doctor recommended a diet rich in seeds, lentils, legumes, and vegetables. Scientists studied the commensal microflora of 305 patients with irritable bowel disease using modern sophisticated techniques of assessment such as Quantitative PCR, cloning, sequencing fluorescence *in situ* hybridization, and electron microscopy and found a high density of mucosal bacteria in sick patients stuck to the intestinal linings. They reported, "*...the density of microbes increased progressively with increasing severity of disease. It is likely that this function can be supported by treatment with a combination of pre- and probiotics (synbiotics).*"

When Alice supplemented with *Quorum Nutriton*™ and followed the diet her doctor recommended, she was able to overcome IBS in just a few months.

Peggy's Hormonal Crisis

The problems caused by taking antibiotics are illustrated in Peggy's case. Peggy was happily married and wanted to have a child. She had taken antibiotics on and off for two years for infections that would not go away. She also took birth control pills for about one year when she was younger. Unable to get pregnant, suffering from bad PMS symptoms, Peggy consulted with her gynecologist. She was diagnosed with large ovarian cysts and told that she could never have a child. Her doctor recommended a complete hysterectomy to minimize cancer risks and stop her PMS symptoms. At 27-years-old, Peggy dreaded the idea of a hysterectomy.

When Peggy asked why she had developed the cysts at such a young age, her doctor could provide no explanations. She was to go back on birth control pills until surgery. Her doctor explained that most women opt for a total hysterectomy to reduce the possible risk of cancer in later years.

Peggy sought the advice of an alternative medical doctor. He performed a series of tests and told her that she was deficient in zinc and other nutrients. He explained that the birth control pills elevated her copper levels and depleted zinc and allowed her to accumulate excessive amounts of toxins that disrupted her hormone balance. Antibiotics caused a leaky gut condition that allowed the entry of toxins and harmful microorganisms into her bloodstream. He also explained that the antibiotics killed off all her commensal cells, causing opportunistic yeast and fungal infections in her gut. For the first time in many years, Peggy had a clear understanding of causes underlying her illness.

Peggy added quorum nutrition to her diet, along with a daily intake of one of the following fats: almonds, sunflower seeds, pumpkin seeds, macadamia nuts, or coldwater fish (cod, mackerel, salmon, haddock or sardines). She eliminated margarine and all saturated fats, and ate lots of vegetables. After 6 weeks of taking quorum nutrition, Peggy felt dramatically better. She returned to her gynecologist for an ultrasound test, and the results indicated that the ovarian cysts were gone. When she told her doctor of her progress on her new diet plan, he insisted that it was the birth control pills that reduced the cysts, not the unique form of quorum nutrition. Peggy had never taken them. Peggy changed gynecologists and had a healthy baby girl only ten months later.

Sadly, the statistics on hysterectomy in the US reveal that many hysterectomies are performed without exploring the underlying causes of dysfunction. Most alternative health care practitioners are keenly aware that the ovaries have the highest concentration of zinc in the female body. They are also aware that copper depletes zinc from the ovaries and kills off commensal cells. Our research has shown that almost 90 percent of females with ovarian cysts are deficient in zinc and commensal cells. When supplementation is given to them to correct these deficiencies, they make incredible progress. However, unknown to most practitioners, the zinc must be in a unique fermented quorum state. I am not talking about amino acid chelates of zinc such as zinc sulfate, zinc aspartate, or other forms of inorganic zinc supplements. I'm talking about zinc formed in a quorum fermentation process that is organically fused to proteins and is instantly absorbed by zinc-deficient cells.

Remember, friendly and beneficial commensal organisms can be easily snuffed out in our bodies. When this happens, commensal cells can't:

- ☐ Work in concert with the immune system to fight off infections;
- ☐ Provide a protective barrier in our gut against the invasion of more harmful microorganisms.

A lack of friendly gut flora allows inflammation-type mediators (chemicals) to quietly chip away at your body day and night. Beyond current nutritional and probiotic products, Quorum Nutrition™ has the potential to turn your life around. You'll be healthier by far and retain that advantage for many decades. You'll feel younger and more vital than you ever thought possible. Supplementing with *Quorum Nutrition*™ may be the single most important change you'll ever make in your life.

As in Alice's and Peggy's cases, our gut ecology can have serious effects on our lives and health. When disrupted it can undermine quality of life and erode human potential. Medical experts now assert that a lack of commensal cells can even contribute to a fatal heart attack.

We are living ecosystems. When you look at hundreds of sick people, all with different diseases and illnesses, you find the common denominator of dysbiosis. Virtually every sick person is suffering from dysbiosis and its resultant non-stop inflammation and immune deficits that cause stubborn yeast and viral infections to dominate the body. In these cases, a quorum has been lost and our living ecosystem can no longer cleanse our bodies of man-made pollutants.

If you are sick, you must see a licensed physician who does not use man-made pharmaceuticals and who respects and wants to work with your inner physician. In my opinion, you must stop consuming man-made chemicals and eat fresh produce. If you don't do these basic things, it's impossible, in my opinion to restore the full power of your inner physician. In the next chapter, you will learn more about how our ecosystems are threatened and extremely challenged in this day and age.

Chapter Three

Understanding Your Body's Living Ecosystem

The hidden physiology of the deep symbiosis between commensal and human cells and their quorum tactics is forcing scientists to see us as living ecosystems. Dr. Jeffry Gordon of Washington University at St. Louis says, "*We're really a composite of species. We have human cells, but there are ten times more microbial (commensal) cells.*" This means that most of our genetic material or DNA is nonhuman. Since commensal cells have hundreds of times more genes that human cells, we can begin to understand why they are the miracle healers of the body.

Stanford University microbiologist David Relman says this internal multitude is like a complex ecosystem—a biosphere, almost. And commensal cells perform some indispensible functions: They help us digest food, produce vitamins and all sorts of living nutrients, and ward off disease. Dr. Gordon calls them "a strategic alliance," a symbiosis between humans and microbial commensal cells that can optimize our performance and health.

Researchers in several countries launched the International Human Microbiome Consortium, an effort to characterize the role of commensals in the human body. With the National Institutes of Health own Human Microbiome Project, new information is bubbling forth on the recognition of commensals and their far-reaching contributions to human health. *"This could be the basis of a whole new way of looking at disease,"* said microbiologist Margaret McFall-Ngai at the 108th General Meeting of the American Society for Microbiology in Boston. *"Human beings are not really individuals; they're communities of organisms,"* says McFall-Ngai. Ironically, all this research is pointing to the fact that the human ingenuity that drives us to understand ourselves is revealing that we're much less "human" than we once thought and that we become a "superorganism" when both human and commensal cells function in symbiosis.

Thousands of scientific studies on the dynamics of quorum communities among plants, insect colonies, and even in human society, reveal a need for symbiosis in nature. Even obesity has a commensal cell component according to researchers in Dr. Jeffrey Gordon's lab at the Washington University School of Medicine in St. Louis.

The Long Hard Road to Discovering Quorum Nutrition

The concept of *Quorum Nutrition*™ grew out of my personal experiences as I searched for an understanding of ADD, in my son Tom's case, and, in my own case of near-fatal kidney disease, neurological deafness and cancer.

In my son's case, doctor after doctor prescribed numerous drugs. Instead of improving, he was made sicker by the side effects of Ritalin and other drugs. In my own case, I still remember the words, "*learn to live with it*," as doctor after doctor diagnosed fatal kidney disease and sent me home to die. I was sent home with no hope for a cure and only one year to live, at the age of twenty. Not one of these doctors was concerned about the fact that I was never breast fed and that I spend most of my childhood on antibiotics and cortisone.

Later, I became a licensed and board certified naturopath and homeopath. My multidisciplinary training gave me a broad perspective and understanding of the multifaceted and complex nature of these disorders. At Louisiana State University I studied neurology, and did research on left-right brain functions. At Baylor University, I studied how the brain regulated balance in the body. And, at the University of Wyoming, I studied how non-stop inflammation was the cause of my allergic conditions and how it had an adverse effect on my brain that resulted in cancer.

The symptoms of brain and neurological dysfunction were not clear cut. For years, I used these symptoms as red flags or clues to find and fix the underlying causes of nerve-related disorders that somehow disabled my inner physician. In saving my own life and stabilizing the brain function of my son, I learned that the pursuit of good health requires looking inward. It means we need a new and deeper understanding of why we are ill or suffer needlessly from stress.

Later on, I found the cause of my disease to be at the quorum level. The body's inner physician or healer is based in the brain and maintained via afferent and efferent neuron reciprocal functions. It has the potential to emancipate you from the need to take pills that spike your energy levels or drugs that suppress your symptoms. It will free you from the way you have been living your life (in ignorance of your inner physician).

But, I was woefully discouraged when every known method of nutrition and botanical medicine failed to stabilize my neurons. I was erroneously aiming my efforts solely at my neurons and human cells. When I asked doctors about the antibiotic damage to my commensal cells, they laughed and said I was crazy. I argued that we're grounded in and aligned by commensal cells, and they are part and parcel of human anatomy.

I had to learn how to ask good questions and make them open-ended and push aside the health myths that were crippling the function of my inner physician. It was about this time in my life that I started studying how bacteria could ferment food and make the kind of nutrients that my deficient commensals cells were supposed to make for my neurons and other cells. When you nourish your cells this way, you can live a life that is filled with balance, energy, and stamina. Part of the reason I'm giving you this new information is so that you'll realize the importance of keeping your body in quorum symbiosis or balance so you can get the ultimate nourishment to your neurons and other specialized cells.

Our inner physician uses nature's handiwork of commensal cells in inexplicably producing spontaneous remissions in many illnesses. Putting these life-saving cells back into our body is as essential for our well-being and growth as sunshine, clean air, and water.

When cell nourishment is poor and stress is high for many years, it takes a unique kind of nourishment to repair our damaged digestive tract and reawaken the full capacity of our neural networks. It must be in a *quorum* nanoscale format and created by nature through the miracle of fermentation. My kidney and neural function recovered with fermented "polar lipids" because they were the exact kind of nourishment needed by my damaged neurons. You'll learn more about this unique kind of nutrition later in this book, and you'll benefit from new ways to experience a higher level of health and wellness.

For over three decades, I saw many patients with ADD, tinnitus, anxiety, memory loss, depression, and other disorders of the nervous system that were all connected to dysbiosis (an imbalance of commensal cells in the gut). I learned the hard way that it was futile to target the brain or organs with nutrition. For nutrition to work and save my life, it had to come from commensal cells. When neurons starve, the afferent brain connections fail to digest our food. When we fail to digest our food, we ferment food into harmful dysbiotic byproducts that feed and encourage the growth of all sorts of pathogens.

The starvation of our neurons from a lack of quorum symbiosis is likely the most common cause of neurological disorders. As I uncovered the startling truths and fallacies of many popular nutritional approaches, I often found myself trying to put together hundreds of tiny pieces in a gigantic jigsaw puzzle—with each discovery adding another piece to the puzzle. In time, the clinical picture of the specific brain nutriture in relationship to commensal cells became clearer and clearer.

Independent of other researchers, I found out how and why neurons were malnourished and developed the concept of *Quorum Nutrition™*. Here are the highlights of a few of my over 350 clinical and scientific studies:

- 1976 in *Journal of the American Audiology Society*, 1:5, 1976 – This study, entitled "Audiologic and metabolic findings in 90 patients with hearing loss," was published in a traditional medical journal. It was the first correlation of diet and faulty gut fermentation patterns as a causative factor in neural deficits of the inner ear. Correlations with diet and metabolism via serial laboratory testing of the blood were significant.

- 1979 in *Audiology and Hearing Education Journal* – This study entitled "Static impedance and aberrant auditory phenomena in 90 patients with cochlear hydrops" documented that dietary deficiencies caused aberrant nerve symptoms.

- 1979 I edited an academic textbook entitled "*Rehabilitation Strategies for Sensorineural Hearing Loss*" published by Grune and Stratton and used in university training programs for over a decade.

- 1981 in *Hearing Instruments Journal* - a study entitled "New Hope for Hearing and Tinnitus Problems: Nutrition and Biochemistry" reveals how attempts to balance body chemistry resulted in improvements in the function of neurons and nerve-type hearing disorders.

- 1982 in the *Journal of Holistic Medicine* and 1983 in *Journal of the International Academy of Preventative Medicine*, I published the exciting results of my clinical research, showing how diet and nutrition can help nerve-type ear disorders.
- 1984 - I edited and wrote three chapters in another academic textbook entitled "*Tinnitus and its Management*" published by Charles C. Thomas, which described the neurochemical basis of nerve transmission and hearing.
- 1988 in the *Journal of Applied Nutrition* 40(2):7584, I published the hospital-based, long-term research project entitled "Dietary and lifestyle influences on cochlear disorders and biochemical status: a 12month study." This landmark study proved that inorganic minerals currently used in the majority of supplements cannot improve the functions of neurons and was not what nature intended in terms of nourishment for neurons.
- 1999 in the *Townsend Letter for Doctors*, I applied my clinical findings to neurodegenerative diseases like Parkinson's and Alzheimer's diseases.
- 2002 and 2007 in the *Townsend Letter for Doctors*, I revealed my new discovery of brain afferent and efferent regulation of inflammation or regeneration.

In research project after research project, case after case, for over thirty-five years, I kept seeing nutrient deficiencies that were not responding to conventional nutrition and patients who suffered in varying degrees from maldigestion. All these patients were acidic—their acid base metabolism was out of balance. Medical physiologists agree that the cells of the brain and body function best at an alkaline pH of around 7.4. In addition, the majority of these individuals had dysbiosis.

When our pH is out of whack, our stomach does not produce acid, our bile becomes acidic, as well as our small intestine, and our large intestine becomes alkaline instead of acidic, making it impossible for commensal cells to find a permanent home in our guts. What many people don't understand is that drinking alkaline water or taking alkaline minerals makes these reversed and abnormal pH gradients in our gut worse.

A comprehensive biological mystery began to emerge as I experimented for decades with trying to mimic what nature intended in terms of probiotic supplementation. But, no commercially available probiotics worked. So I had to invent an 8-strain blend that could unfold with proper quorum-based nourishment into a symbiotic community of cells that nature intended before we ever took an antibiotic.

Quorum Nutrition™ was born out of my strong desire to share with others the crucial revelations about nature's wisdom that changed my life for the better. Throughout this book, I have relied on observations, analyses, scientific fact and experiment, and the deductions that can logically be derived to support my statements.

The sum of evidence about our inner physician, I believe, presents an emergence of a human *superpotential* or quorum superhealing that can be activated and brought into harmony with every facet of our physiology. Tapping into the hidden secrets of life-governing functions and automated self-healing functions of the body will add measurably to the quality of your life. What follows in this book is the result of three decades of research aimed at developing a dynamic, quorum method of actually transforming and rejuvenating our bodies.

Years of trial and error research and multi-disciplinary training helped me to discover how and why commensal cells acted like a strong battery connection to the brain's inner physician. Fixing its dysfunction was the key that unlocked the door to releasing huge amounts of nature's healing energy throughout my body.

On the edge of new discoveries about the laws of nature, the true nature of healing, and the untapped sources of the body, I founded the *American Academy of Quantum Medicine* to board-certify and train doctors in a form of healing that required minimal practitioner intervention. Rather than search for relief of their symptoms, patients seeking help from Quantum Medicine™ practitioners were learning how to take control of their own inner physician and their self-healing abilities.

During the development of this new field of medicine, it became evident that many of the world's greatest and time-tested healing methods were formulated at a time when the world was not as toxic or as stressful as it is today. Hundreds and thousands of years ago, when Traditional Chinese Medicine (TCM) and Ayurveda were formulated, healers couldn't imagine the negative impact of living in today's polluted and stress-driven world.

The Commensal Cell Connection to Disease

Commensal cells stake claims in our digestive and respiratory tracts, our teeth, our skin and nearly every part of the body. With quorum sensing, they establish increasingly complex communities that may be likened to a forest that gradually takes over a clearing. Our bodies harbor 100 trillion commensal cells, outnumbering our human cells 10 to one. It's easy to ignore this astonishing fact because these cells are invisible to us. We have been conditioned to disinfect everything and to attack and destroy both good and bad microbes in our body.

Dr. Jeremy Nicholson of Imperial College London found that commensal microflora *"help us absorb nutrients and fight off viruses and bad bacteria; disrupting intestinal colonies, such as with a course of antibiotics, often leads to digestive sickness. In fact, almost every sort of disease has a gut bug connection somewhere."*

Culturing our commensal microflora has been a challenge to scientists because it's impossible to culture stool samples as they survive only in highly acidic, oxygen-free environments. But, new DNA-sequencing technologies have scientists starting to identify the strains of gut bacteria, and there is growing interest in doing so: The National Institutes of Health launched its Human Microbiome Project last year with the goal of fully characterizing the commensal cell microflora.

Double-blind, placebo-controlled trials, have shown that restoring colonies of commensals can effectively treat ulcerative colitis and irritable bowel syndrome. *"It opens up visions of a future that we would never have suspected even a few years ago,"* Nicholson says. "*Many microbiologists might argue this is fanciful, but you only make huge progress in science by thinking almost the unthinkable*."

Quorum commensal communities are examples of self-sacrifice for the benefit of the larger colony and human cells. They form physically close communities in which some cells exist solely to provide structural support or protection for others. Symbiosis shows us how all our cells are living together in a continuous, interconnected way. For example, when scientists discovered life in the deep-sea hydrothermal vents, they could not explain how gigantic tubeworms could live in scalding-hot water filled with hydrogen sulfide. Later they discovered that these worms harbored symbiotic bacteria that could eat hydrogen sulfide and turn it into something usable by other life forms. The discovery underscored the fact that nature employs commensal cell communities of microbes to maintain all sorts of ecosystems.

We are multisensory and multidimensional beings entwined with nature. Impulses, emotions, and intuition come to us through all senses and originate from cellular quorum states or symbiosis and our inner physician. To be truly healthy, we have to honor this design in nature and increase our awareness of how it works on a quorum level. This book is designed to help you grow in perceptual strength and understanding of the complexity of this awesome intelligence or intuition.

Quorum Nutrition™ made by commensal cells should be an integral part of anyone's immune nourishment program. Quorum nutrients are not like the vitamins and minerals in your daily multiple vitamins. They are raw, food-derived, and fermented or pre-digested. A lack of these quorum nutrients can corrupt the quorum code and jam up cell-to-cell communication. For this reason, scientists are putting less emphasis on the Genome Project (identifying the genes) to the Human Microbiome Project.

Quorum Nutrition™ helps to keep our immune system watchful and strong enough to immediately respond to invading and threatening infections. When commensal cells are nourished this way, they can strengthen immune responses to better combat infections and boost repair and regeneration. Nature certainly knows what she's doing in the case of breast milk, which contains quorum nutrition—a fact that in and of itself supports the notion that the body needs this kind of nutrition to stay healthy.

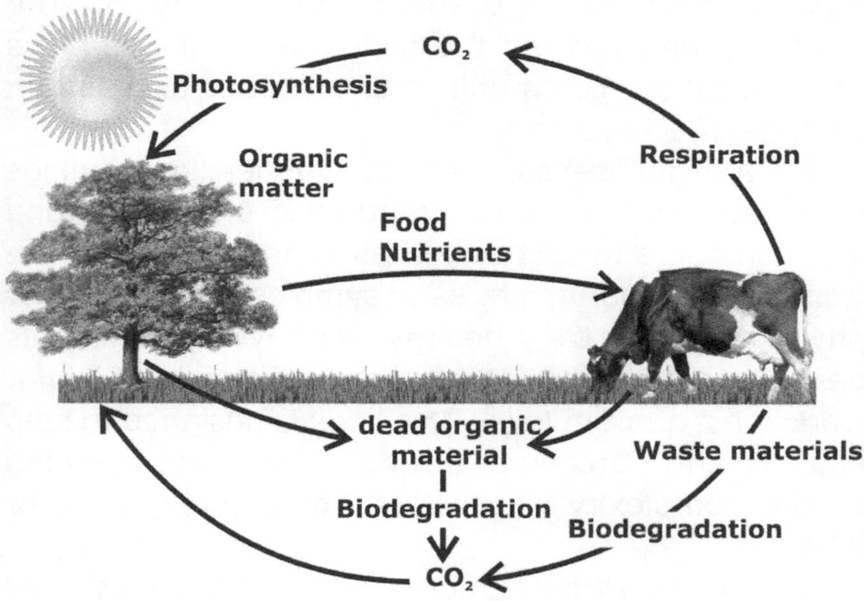

In the 2008 journal *Nature Biotechnology*, it was stated that only 1 percent of microbes survive in the laboratory, and the remaining 99 percent are undiscovered in nature's ecological cycles that cleanse the earth. Dr. Chistoserdova estimates that one mud sample contained about 5,000 different microbes. "*These are the bacteria that maintain the Earth's health. Some of the methane escapes -- in some parts of the lake you can see the bubbles. But whatever doesn't escape as bubbles, these bacteria do a very good job of eating it,*" Dr. Chistoserdova said.

Researchers at the University of Wisconsin-Madison explained nature'sCarbon Cycle. "*Organic matter* (*CH_2O*) *derived from photosynthesis (plants, algae and cyanobacteria) provides nutrition for animals and associated bacteria which convert it back to CO_2. Organic wastes, as well as dead organic matter in the soil and water, are ultimately broken down to CO_2 by microbial processes of biodegradation*."

Intuition is perception beyond the physical senses that is meant to assist you. It starts with the symbiosis of cells and extends itself to the deepest core of our being. But, keep in mind that if you ever took an antibiotic or have waged war on microbes with anti-infective herbs or food concentrates, you are in dysbiosis, the opposite of symbiosis.

By wiping out commensal cells, we short-circuit the awesome wisdom and intelligence of our inner physician. When we are out of balance with nature, this energy dwindles when we overstress or over-stimulate our bodies and minds. And, we depress it even further by reaching for symptomatic relief.

Stress can get out of hand (stress versus distress), when we are toxic or nutrient-depleted. But, when we live in harmony with nature, this amazing intelligence is able to make us stress-resilient and repair stress-induced damage to our bodies at lightening speeds. It is able to boost our immune system and keep us free from pathogens. And, you are held responsible for how you use it or misuse it. Ideally, you want to take this knowledge and use it to benefit yourself and others.

So powerful is the energy from innate intelligence that it's responsible for the physical template that your entire body grew into from a mere sperm-egg interaction. Right after conception, it formed all the different systems of your physical body. How can we not trust this intelligence? It is this innate intelligence that empowers us.

What we need to understand is that these intuitive and innate healing processes are contaminated by our:

1. Failure to nourish and restore all the cells in our body.

2. Failure to implant or colonize our bodies with commensal cell microflora, and

3. Use of, and accumulation of, too many synthetic, man-made chemicals, toxins, and electromagnetic stressors.

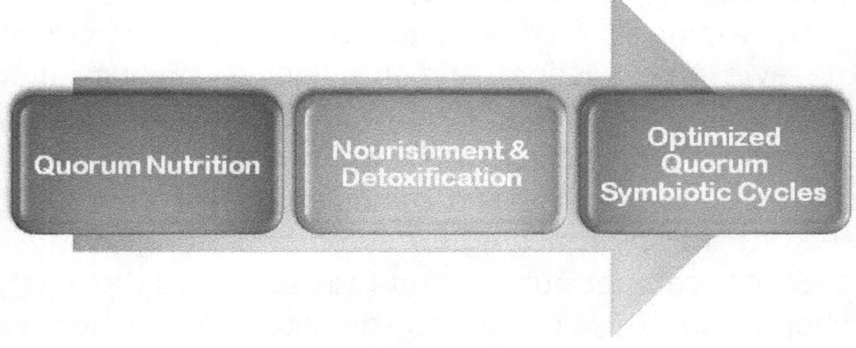

In over 30 years of watching the successes and failures of every system of natural healing and working with thousands of patients over the years, I can say with some authority that the core of illness is dysbiosis and a disruption of quorum cell-to-cell communication that weakens our afferent regenerating neurons. Dysbiosis occurs in the gut at the abdominal center or crossroads, called the *Chung Mo* in Oriental Medicine, where deep energy channels (acupuncture meridians) intersect with the nervous, lymphatic, circulatory, and digestive systems.

Years of studying this energy flow have taught me that the primary reason humans don't respond to most treatments relates to dysbiotic generated energy blockages in the abdomen. Feel your body in the area of the abdomen while lying face up. Notice if there is any tightness or pain, or other sensations. You don't have to be a doctor to realize that tense or tender areas reflect too much inflammation and excessive efferent neuron tissue injury or degeneration. When you damage your digestive organs, diet doesn't matter as no matter how good the food is that you eat, you will not digest and derive nourishment from it.

As you become more aware of these problematic areas in the abdomen, you will also become more aware of the progress you make when you get closer to nature's way of healing and nourishing the body. You will actually feel the abdomen become less tense and tender in areas where it was sensitive before. Illustrations in the next few chapters will reveal how brain afferent neurons regulate digestion, detoxification, hormone balance, and immunologic functions. To get the proper nourishment needed by the afferent neurons that regenerate the body, you have to implant an *8-strain Quorum Commensal Probiotic Colonizer* in your gut, and this is explained in detail in the next chapter.

Chapter Four

Restoring Your Body's Living Ecosystem

There's a growing consensus among scientists that the relationship between human and commensal cells is much more of a two-way street. With new technologies that allow scientists to better identify and study the commensal cells that live in and on us, we've become aware that commensal are powerful

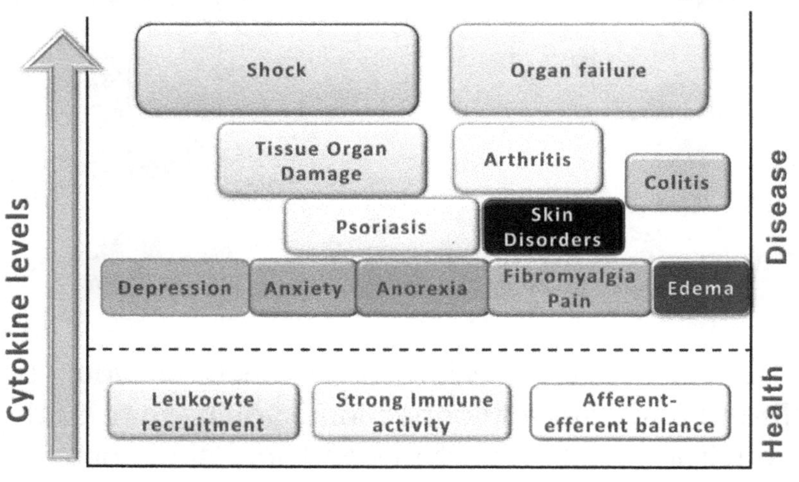

Immunology Today 18:17:1997; *Current Opinion in Immunology*, 3:5, *1991; Immunology Today*, *18:7, 1997: Journal of Neuroimmunology*, 32: 3, *1991*

nutrient and chemical factories that affect how human cells and neurons function.

Commensal cells die and become toxic from antibiotics or man-made chemicals. The man-made chemicals immediately induce inflammation, triggering brain efferent neurons to pump out anti-inflammatory cytokines, that over time, can injure our human cells. Studies in prestigious immunology journals have shown that elevated cytokines cause disease.

When the efferent neurons go up, the afferent regenerating neurons used by the inner physician go down. This see-saw action explains why so many sick people have maldigestion, toxicity, and immunological dysfunction as afferent neuron function becomes deficient relative to efferent function.

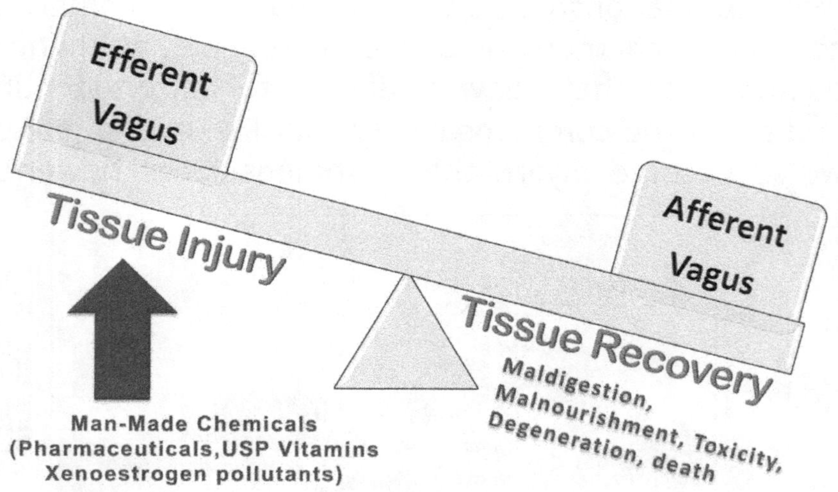

To restore the reciprocal balance of your brain efferent and afferent connections, you need to cultivate an environment that nurtures and nourishes commensal cells. And, too, maximizing the commensal growth and proliferation will secure their ecological niche and start empowering your inner physician.

If we do not put back what was destroyed by antibiotics or pollutants, we will make more visits to physicians and be more vulnerable to infectious diseases, even fatal ones. Reciprocity (division of human and commensal cell functions) is a tool that nature uses so our cells can enhance our being through cooperation—working for the common good of the entire body.

To accomplish this goal, you have to reenergize your commensal cells by restoring coalitions and alliances with the human immune system. If you do not do this, then you will always be chasing symptoms and never really treat the cause of your sickness.

Now that scientists finally acknowledge that most of our cells aren't human, we need to focus on restoring commensals that are deficient in our bodies. Sadly, as I explained in my own journey to wellness none of the common probiotics sold in the marketplace can replace them. My 8-strain Quorum Commensal Probiotic Colonizer is able to closely mimic what nature originally intended to dwell in our bodies and what, with proper long-term nourishment, can unfold into hundreds of very specialized commensal cells. Doctors that are board-certified in Quantum Medicine are using these discoveries with amazing clinical success (see Appendix).

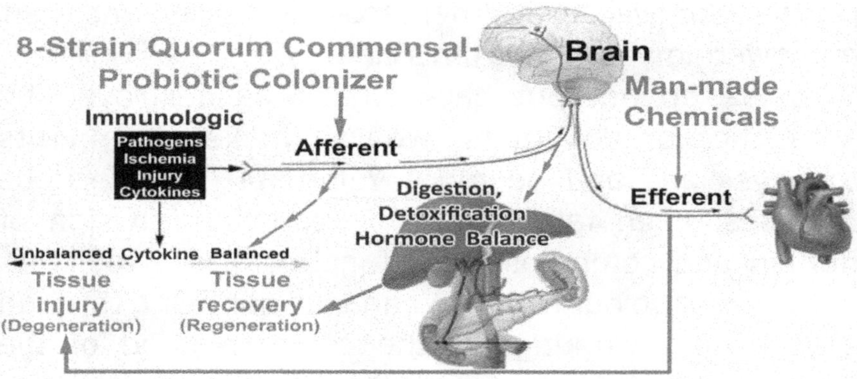

Just like a battery needs the bi-polar *positive* and *negative* polarities to work, neurons function best when we cleanse the body of man-made chemicals. All man-made chemicals have a positive ionic charge with no negative charge. Try running your car on a battery with just the positive connection, and you will see what I mean.

Antibiotics that we down at the first sign of a cold or flu are waging an unjustified civil war against our commensal cells. Add to that the massive quantities of antibiotics we feed to our livestock antibiotics which ultimately end up in our bodies, disrupting quorum symbiosis and our body's delicate ecological balance.

Dr. William Karasov, a physiologist and ecologist at University of Wisconsin–Madison, says, "*We've all been trained to think of ourselves as human. Bacteria have been considered only as the source of infections, or as something benign living in the body*." Now, "*We are so interconnected with our microbes that anything studied before could have a microbial component that we hadn't thought about*."

When appreciating the intelligence of your inner physician, there's no stronger evidence than to trace back how human life proliferated and differentiated from conception. With motility or oscillatory motion in the direction of our original embryologic migration & origin and the inclusion of commensal microflora that are colonized and nurtured, we can restore the function of our living ecosystem and health.

Change—or should I say lasting change—requires respecting and listening to the subtle messages of the inner physician. Each time you reach for caffeine, sugar, or a B vitamin product to remedy fatigue, you turn off its healing power. Each time you reach for a supplement made by nature's quorum fermentation process, you can turn it on.

The quick-fix stimulation approach only drains the voltage or battery-type charges needed by your inner physician to heal your body. Research shows that when this polarity is disrupted, our hepatocytes (liver cells) fail to produce bile. Without ample bile, we cannot get rid of man-made chemicals that are sequestered deep in our tissues and organs. If you've always relied on food or beverages to spike your energy, you will have to get used to the idea of turning on this energy the natural and healthy way.

Sometimes these quick-fix addictive tendencies come from being overly stressed, angry, or unhappy with life. In these cases, you've got to figure out why you need ice-cream, coffee, or soda to feel better. What in your life is making you fill this void in energy with foods and beverages that turn off innate healing?

The innate power of our inner physician is unknowingly suppressed by nearly everything we do in our daily lives. The supplements we choose to take, the food we eat, and the way we think and respond to the stress of modern day living and even electropollution can drain and zap this healing energy.

The Hidden Health Dangers of Electropollution

We live and are surrounded by a sea of electropollution. Unknown to most people, electromagnetic frequencies (EMFs) generated from satellite frequencies (cell phones, TVs, wireless technology, etc.) suppress inner physician function by causing interference in quorum cell-to-cell communication.

EMF overload in your body is like a blown fuse in your home that stops the flow of electricity. Likewise, the electrical energy flowing through the neurons that control the digestive process in the *Quorum Symbiotic Cycle* can get disrupted. Since EMFs have doubled in the past two years, it is important that we understand how to protect the body from them. And, in the appendix of this book, you'll be given helpful guidelines regarding how to insulate and fortify your inner physician from the constant interference of EMFs.

EMFs from electrical equipment and common household and office devices can continually throw off our bodies' equilibrium. They also upset the brain/nervous system's control over tissues, organs, and systems of the body. Since the nervous system uses simple, nonlinear electrical patterns to maintain homeostasis, it is extremely sensitive to environmental frequencies that are similarly simple and nonlinear.

Concern over health effects from EMFs was prompted as long ago as 1979, when a 7 million dollar study by the California Department of Health Services showed how children who lived in close proximity to electrical lines had a greater than 50 percent probability of developing childhood leukemia, as well as risks for adult brain cancer and Lou Gehrig's disease. In addition, it suggested a 10 to 50 percent possibility that EMFs could increase the risk of male breast cancer, childhood brain cancer, suicide, Alzheimer's disease, and "sudden death." Here's the good news: There are a number of devices in the Appendix of this book that will give you daily protection from EMFs. **

Today, EMFs are more dangerous. Satellite-controlled cellular phones and television and computer satellite links permeate the air and our bodies with coherent high frequency energies. These high frequency EMFs permeate our bodies, adding more information to the interaction between innate intelligence and the cells of the body. The effect is like trying to listen to a radio station while several stations' signals vie for the same narrow strip of the band. The result is static with only intermittent clarity. That's why I invented a device to protect against overload from satellite-generated high frequencies. You can learn more about this device later in this book. **

Ecosystem Stressors

As I continued my research, it became obvious that correcting or balancing human and commensal cell balance was not enough. I had to find ways to free up, strengthen and harmonize commensals from a hostile environment. Consistently, I found that commensals could not be restored without daily supplementation of quorum nutrition™.

The vulnerability or resilience to the stress of life in my patients was clearly related to the status of the body's ecosystem. Stress overload with man-made chemicals has an effect similar to water slowly leaking out of a battery, causing the battery to get weaker and weaker. Thus, I viewed my therapy, not as a direct objective treatment of disease and dysfunction, but rather as a prod to the inner physician to restore the balance of brain efferent-afferent nerve flow and quorum symbiosis. Maximizing the genius of your inner physician maximizes your own life and secures the ecological niche in your most important commensal cell arena.

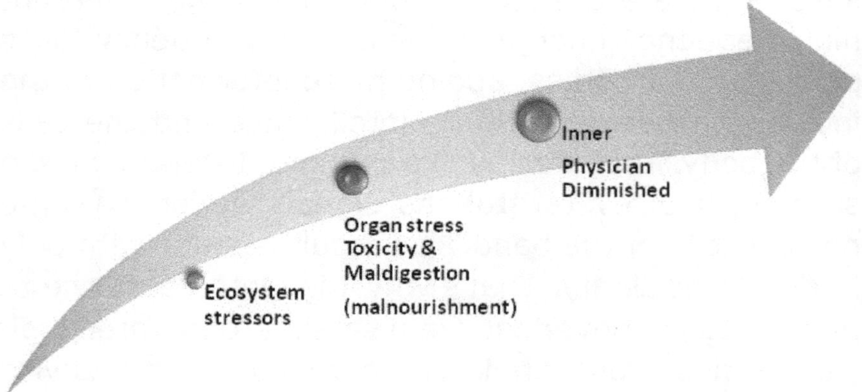

In my research, both university-based and clinical, I perceived myself as an educator rather than a healer that guided my patients to make contact and restore their inner physician. In case after case, I would align the patient's focus on the abdominal tensions caused by dysbiosis. Prodding these deeper neural circuits and pathways brought patients into deep contact with their own ecosystem abnormalities. They would learn, as I did, that their body was not whole and maladapting to stress.

At my board certification seminars, I would train doctors how to assess the inner physician dynamics and its "defensive armor" against stress that blocked innate healing. In every case, these hidden blocks were manifested as deep areas of abdominal tension or tenderness upon palpation which, in turn, caused deep visceral (organ) tensions that were traced back to toxic and starving afferent neurons.

Teaching doctors the neuromuscular patterns of inner physician dysfunction provided them with a natural antidote to ecosystem stressors and proved to be the most revealing measure that would help them bring their patients into wholeness and harmony with their environment. Their health improvements were attributed to instant cell quorum nourishment™ that bypassed maldigestive tendencies and quickly nourished the commensal cells and neurons.

The most exciting effect was that they were able to get their patient's to focus on the particular aspect of their lifestyle that was enhancing or diminishing their inner healing capacities. Overstimulation from addictive foods, sugars, caffeine, nutritional supplements, and EMFs proved to be their worst enemy. By removing these stressors and giving them instant cell nourishment via *Quorum Nutrition™*, the latent and little-used ability of their body to heal itself seemed to become more operational.

Megavitamins and other man-made nutraceuticals or drugs (pharmaceuticals) drastically suppress our healing energies, making us more susceptible to assaults from stressors, whether physical, chemical, or emotional. In following nature's wisdom, we have to realize that nature never put vitamin C or vitamin B in a food by itself. Likewise, isolating antioxidants out of a whole food does not follow the path of nature's wisdom.

Stimulation from commonly used magnetic, microcurrent, low level laser, LED or EMF-generating devices also goes against nature and seems to prevent the inner physician from fully expressing itself. As a result, people go through life accepting stress and disease and not compensating for the damaging blows of stress without realizing that their actual disease is caused by suppression of inner physician-guided healing.

Despite having the world's finest medical care for treating traumatic injury, preventing disease or activating the body's innate healing powers is sadly neglected in the USA. With no training in Quantum Medicine, French acupuncture, or other time-proven, natural healing methods that seek to reestablish the body's own energetic balance and vitality, they commonly use pharmaceutical or nutraceutical treatments that do not cure the condition. Instead, they simply block symptoms. When you block a symptom with a drug or a synthetic vitamin, you block innate healing. A blocked symptom often rebels by spawning clusters of symptoms that doctors diagnose as disease or syndromes. Then they "treat" the disease by just handing out a prescription for every symptom. Shouldn't doctors be educating patients about the causes that underlie their symptoms? I say, yes, they should. Shouldn't they teach their patients how to take greater responsibility for their own health? I believe physicians have a responsibility to do so. That's one reason I have written this book.

In order to facilitate real healing, it is necessary to understand why and how quorum symbiosis needs daily nurturing to restore itself in today's polluted world. Man-made chemicals are all undercurrents that pull us away from inner physician function. We have to come to grips with whatever is underlying it, and restore the body's living ecosystem.

This is exactly the opposite of what mainstream medicine teaches and practices. However, more and more doctors are focusing on prevention and on caring for the body in such a way that its natural, self-sustaining abilities are enhanced. Brutal diseases like cancer, diabetes, heart disease, and HIV kill millions of people, and a new paradigm is desperately needed.

Since the general consensus among medical scientists is that only about 15 percent of modern medicine is proven scientifically, it is a lost cause for them to demand scientific proof of nature. In fact, a 1997 *Journal of the American Medical Association* article reported that the likelihood of finding any specific disease on which to pin a patient's symptoms was less than 2 percent. Whether they admit it or not, modern medicine has been shocked by the rich consequences of changing lifestyles on human health. The *American Cancer Society* and other leading medical organizations now advocate a healthier diet as a key component of preventing serious degenerative disease in the USA.

Some of the most widely accepted academic teachings, intellectual beliefs, and therapeutic modalities often make the problem of being sick—from any cause—many times worse than it has to be. Hippocrates, the founder of modern medicine, made little use of drugs, believing that they simply interfered with the natural healing processes. He used "fomentations" (botanical medicine compresses), bathing, diet, and other natural remedies to detoxify and nourish the body.

Modern medical treatments frequently, in effect, stress the body, underestimating its magnificent intelligence and blocking its capabilities for self-healing and repair. However, a growing number of frontier scientists, some of them Nobel Laureates, have been assiduously studying the amazing health benefits of our commensals. Now, this work has progressed to a point where real solutions are available. The following safe, proven, and time-tested natural therapies can be used to evoke inner physician healing, even for the most stubborn cases of a wide spectrum of ailments:

- An exciting new form of nutrition, called *Quorum Nutrition™* that's utilized by your human and commensal cells with great efficiency and that can flood your body with healing energies and nutrients**

- A new form of revitalized and energized water designed to hydrate and cleanse the body.**

- A wearable pendant that works in harmony with your inner physician, while shielding your body from the chaotic interferences of EMFs (including harmful satellite frequencies, which stress and inhibit our innate healing energies)**

- A diet-lifestyle plan designed to remove stressors to your body's inner healer and activate extraordinary healing energies.**

These healing breakthroughs involve unleashing the awesome healing power of your entire body, allowing it to function the way it was designed to function, so it can heal itself naturally.

Many scientists believe the secrets of regeneration and healing lie not within costly drugs or expensive medical treatments, but in the body's own innate intelligence and afferent neuron regulated regeneration. Hence, finding ways to nourish and fortify afferent neurons will bring you long-term health.

This nutritional breakthrough changes nutrient delivery as we know it and gives us a way to rapidly correct nutritional deficiencies caused by years of stress. Imagine: eating easy-to-assimilate and fast-acting food nutriture to tame the ravages of everyday stress!

When using nature's timeless wisdom and probiotic counterparts, the body is nourished 300 times faster than eating whole foods or taking ordinary supplements. Nature's most powerful healing agents are quickly propelled into cells to fortify stress defense mechanisms and to insure that nerve energy keeps flowing despite everyday stress encounters.

Rather than function like lifeguards who rescue patients from one health disaster after another in the dangerous surf of today's toxic and stressful environment, doctors of the future will ultimately teach us how to stay healthy by avoiding health stressors and by enhancing and nourishing the body's intelligent, life-sustaining energy field.

If you are suffering from an unresolved illness, your body's inner physician is blocked, stalled, or overwhelmed by stressors, and it needs help. Supplementing with *Quorum Nutrition*™ provides instant utilizable nourishment that helps to re-direct blocked or stagnant energies into stress-damaged areas for self-healing to take place optimally.

New self-help tools include a revitalized and energized water to free up your constricted energies and a wearable energizing pendant that shields your body from the harmful effects of cells phones, computers, and EMFs (see Appendix). Restoring your living ecosystem as nature intended will provide you with a strong ally in the fight against stress (see Appendix and Resource section for more information).

Chapter Five

Why We Need Quorum Nutrition

One hundred years ago, life in the country was full of fresh air, clean water, and rich soil. The energy of the sun and the earth's magnetic frequencies were naturally resonant. Now, however, due to innumerable factors, including overpopulation, deforestation, and personal and corporate pollution, this environment has slowly been degraded and contaminated.

Our bodies have to deal daily with a huge array of pollutants. When environmental toxins build up in the body, they cause fatigue, weight gain, allergies, headaches, skin eruptions, slower mental processes, and a weakened immune system.

Detoxification involves more than taking saunas or fasting. The process of detoxification should be an ongoing daily process. There is mounting evidence that human exposure to chemicals, even at low levels, can be harmful. Such exposure is linked with adverse biological effects, including endocrine disruption, chemical sensitivity, delayed healing, prolonged infections, and cancer. Lifetime toxic exposures can cause irreversible injury to the body.

According to world-renowned authority on quantum physics, Dr. Fritz-Albert Popp, toxic chemicals act as biophoton light scramblers. During wartime, communications are kept from an enemy by using devices designed to scramble messages, thereby creating confusion and misinformation. Biophoton scramblers act in an identical manner, causing DNA instability and disrupting quorum cell-to-cell communication. While normally quite adaptable and resilient, our bodies are also perhaps the most delicately tuned "machines" on earth. Their intricate functioning is affected not only by lifestyle stresses and the foods we eat, but also, both positively and negatively, by numerous factors and substances in the environment.

Changes in the environment, in available foods, in nutritional supplements (over 96% of supplements are toxic or irradiated), and even in medicines are all combining to bring about extremely stressful living conditions. According to the *National Institute of Environmental Health Sciences*, more than 300 billion pounds of synthetic chemicals were produced, used, and disposed of in the early 1990s. Almost 80,000 chemicals are registered for commercial use, with 2,000 additional new ones being added annually to food, personal care products, supplements, prescription drugs, household cleaners, and lawn-care products.

Pollutants are being stored, rather than neutralized and excreted out of the body, thereby depleting our bodies' nutrient and self-healing reserves at an alarming rate. Many carcinogens or potential carcinogens enter the bloodstream without being detoxified. Un-

fortunately, fasting and common homeopathic and nutritional detoxification strategies may allow these un-detoxified toxins to circulate and do further damage to the organism. Rather than pursuing aggressive detoxification strategies or fasting, consuming Quorum Nutrition™ allows these toxins to safely leave the body so the innate healing mechanisms of the inner physician are not disrupted. As long as toxins remain in the body, our innate healing system functions at half-potential.

Body Burden and Our Chemical Legacy

As medical researchers are documenting, toxins from household chemicals, industrial pollutants, food additives, and pesticides interfere with almost every facet of human functioning. One study, done at Mount Sinai School of Medicine in New York, reported an average of 91 industrial compounds, pollutants, and toxic chemicals in the blood and urine of nine volunteer test subjects. These toxins in the body are cumulative and are called a person's "body burden." Of the 167 toxins isolated from the bodily fluids of these test subjects, 76 are known to cause cancer in humans and animals, 94 are toxic to the brain, and 79 can cause birth defects.

Confirmation of human toxicity comes from the U.S. government's *Centers for Disease Control and*

Prevention (CDC). The CDC reported that of 116 environmental toxins used in consumer products, 89 were positive in tests of people's blood and urine. According to this report, PCBs, dioxins, organophosphate pesticides, herbicides, insecticides, and synthetic disinfectants were revealed by the lab tests to be present in these body fluids.

Today, our body burden reflects our chemical legacy, that is, over 50 years of increasing reliance on synthetic chemicals for every facet of our daily lives.

The latest report from the CDC: We are all carrying around toxic metals, plasticizers (phthalates), combustion products (polycyclic aromatic hydrocarbon metabolites and dioxins/furans), chlorinated hydrocarbon pesticides (DDT, hexachlorobenzene, lindane, chlordane, heptachlor epoxide), PCBs, organophosphorus insecticide metabolites, other pesticides (various herbicides, including atrazine, 2,4-D/2,4,5-T and pentachlorophenol), tobacco smoke indicators (cotinine) and xenoestrogens. Sadly, all these toxins have been implicated in dramatically increasing the incidence of cancer and other diseases. No wonder only 1 out of 15 people died of cancer in 1970 and today almost one 1 of 2 people ends up with cancer!

Mothers pass on hundreds of chemicals - from pesticides to flame retardants - to their babies through the umbilical cord, according to a groundbreaking study. The study found 287 industrial chemicals and pollutants in umbilical cord blood from 10 babies born in U.S. hospitals. The blood harbored pesticides, chemicals from non-stick cooking pans and plastic wrap, long-banned PCBs and wastes from burning coal, gasoline, and garbage. On average, each baby had been exposed to 200 chemicals while in the womb.

The *Environmental Protection Agency* (EPA) found that asthma in children has more than doubled since 1980 (something the report connects to air pollution) and that mercury exposure has increased dramatically. According to the EPA report, about 8 percent of pregnant women have enough mercury in their bodies to significantly increase the risk of fine motor problems, poor language, poor memory, and poor visual-spatial skills in their children. As little as one part per billion of methyl mercury has been shown to cause harm. Sadly, about half of today's young women have this amount or more in their bodies – and transfer it to their babies during pregnancy.

"*The numbers are startling when you hear them*," said Dr. Alan Greene, a *Stanford University* pediatrician. In the month leading up to the baby's birth, the umbilical cord pumps at least 300 quarts of blood each day back and forth from the placenta to the fetus, bringing the baby oxygen and nutrients. Scientists once mistakenly thought the placenta shielded cord blood and the baby from most chemicals and pollutants. Dr. Greene described the placenta as a "free-flowing, living lake" that the blood vessels in the umbilical cord draw from. "*Today, this most primal of lakes has become polluted with industrial contaminants*," Greene said. "*And developing babies are nourished exclusively from this polluted pool. They mainline the contaminants through their umbilical cord, injecting them into their veins more potently than any IV drug administration*."

Since current federal government regulations on pesticides do not adequately protect children, nourishing ourselves and our children with *Quorum Nutrition*™ can help to fortify our innate detoxification abilities to excrete these dangerous toxins safely out of the body. Opening up the deeper energetic zones that carry the chaotic and foreign homeopathic imprints of these toxins is critical for innate healing to be fully operational.

Environmental contaminants can affect children more aggressively and cause serious harm. A growing baby does not have a mature immune system and large enough organs to excrete these body burdens. Even adults need to be concerned, as toxins they acquired since birth may still be locked in their systems, causing a wide spectrum of weak immune responses and other symptoms.

Today, it is impossible to be in a public place and not inhale hundreds of toxic chemicals. Even our food and water supplies have become loaded with these toxins. A high percentage of household and body care products and fragrances (perfume and cologne) are loaded with carcinogenic toxins. Paints and varnishes, gasoline, glues, clothes dry-cleaned with solvents, plastic food containers, and home and garden pesticides are other common sources of toxins. The result: a long list of health problems that includes direct damage to the lungs, liver, kidney, bones, blood, brain, and nerves, and the reproductive system. Hundreds of adverse health effects can arise from exposures to chemicals or metals and include cancer; high blood pressure; asthma; deficits in attention, memory, learning, and IQ; Parkinson's-like diseases; infertility; shortened lactation; endometriosis; genital malformation; peripheral nerve damage; and dysfunctional immune systems. For example, dioxin is a carcinogen, and fetal exposures to dioxin interfere with normal development, including the development of the immune system.

Fetal exposure to polychlorinated biphenyls (PCBs) is related to behavioral and cognition problems. DDT exposure has been related to women's inability to produce sufficient breast milk. The immune systems of children are unable to produce enough antibodies to survive the increasing number of lethal and mutant viruses.

Nobel Laureate research confirms the need for proteins to carry information like "zip codes" for nutrient transport and utilization by our cells. One of the most remarkable advantages of *Quorum Nutrition*™ is the relative ease with which the proteins can be joined to their inherent counterparts (minerals, enzymes, etc.) to detoxify the body safely via the liver. Unknown to many nutritionists, isolated amino acids and synthetic vitamins (found in over 95% of all supplements today) jam cellular signals and serve no function in liver detoxification pathways. *Quorum Nutrition*™ gives proteins an amazingly strong ability to chelate or attract toxins and carry them safely out of the body. This process allows proteins to be rigged naturally to tow toxic payloads directly out of the body. Fast clearance by binding to toxins, pathogens, or aberrant signals—offered by nature—is evident in *Quorum Nutrition*™.

In review, our inner physician gets short-circuited when neurons become impregnated with these toxins. Efferents soar and afferents fail. The full flow neuron or nerve transmission falters, causing the nervous system to perform inefficiently and in compensatory modes of survival physiology. Remember, the body is an omnivorous harvester of information. Its cell receptors can vibrate open, like a key that opens a lock, or stay shut when the zip code quorum language of the cell is missing.

Common food additives, preservatives, and packaging and processing methods used by the food and nutritional industry have increased one's need for antioxidant protection dramatically. Indeed, studies have shown that our modern day diet accounts for 35 to 80 percent of cancers (*JANA. 2000; 3(3); Nutrition Research 1996; 16; Journal of National Cancer Institute 1981; 66; World Cancer Research Fund. Washington D.C. American Institute for Cancer Research, 1997; British Medical Journal 1998; 317*).

To make matters worse, 97.6% of nutritional and herbal supplements were found to be toxic in a University of California study (*Journal of American Nutraceutical Association*, 1996, 2:1). Despite evidence that ionic, inorganic or non-covalent minerals are toxic and carcinogenic, they are in nearly all supplements today (*June 2001. Townsend Letter for Doctors June 2001, December 2000; Nutrition & Cancer 1992;18; Nord. Medicine 1982 111(7); Mutation Research 1996, 370:1; Journal of Environmental Pathology Toxicolology 1980. 4; Nutrition & Cancer. 1983; 4*). One has to wonder how anyone can think that the inorganic iron in a rusty nail can be superior to the organic iron in a raisin or molasses.

The *American Cancer Society*, The *American Heart Association*, and *the National Cancer Institute* recommend 7 to 9 servings of fresh fruit, vegetables and grains daily to significantly reduce the risk of deadly diseases. Yet, because stress and the aging process deplete digestive capacity dramatically, and mass manufacture and distribution of processed foods with poor handling, storage practices, and methods of food preparation degrade the plant's critical, health-promoting nutrients, supplementation is necessary.

Since food additives/synthetics are almost ubiquitous in manufactured supplements, foods, and drinks, more Americans are consuming organic foods. While organic foods are richer than commercially-grown food in nutrients, many organic farms are located in highly polluted areas and/or use contaminated water to irrigate their crops. This means that significant levels of carcinogens are in organic foods from the infiltration of the soil with acid and pollutant-laden rain. The widespread use of food irradiation is another source of contamination of our food supply.

Avoid Toxic Environmental Estrogens

Estrogen, naturally present in all of us, is a hormone that regulates many physiological processes, including growth and reproduction. An environmental estrogen, sometimes called a xenoestrogen (literally a "foreign estrogen," and also referred to as endocrine-disruptor, endocrine-modulator, ecoestrogen, and hormone-related toxicant) is a substance that acts like an estrogen hormone in living organisms. It blocks the fat-burning thyroid hormone and causes the body to store fat excessively. Exposure to environmental estrogens occurs throughout our lives, both from the food we eat, and also from household products, including detergents, drugs, aerosols, lubricants, dyes, cosmetics, pesticides, synthetic fabrics, and plastics. Direct exposure also comes from drinking water that has been contaminated by chemicals and their toxic byproducts released into it by industrial discharge and sewage waste. We are also indirectly exposed when chemicals are released into the air and water, and when airborne fumes from industry or hazardous waste incinerators land on grass or hay and are eaten by livestock that is subsequently consumed by humans.

Many of these xenoestrogens have been associated with developmental, reproductive, and other health problems in living organisms. And since they are stored in body fat cells, they adversely affect fetal development in a mother's womb and, later, infant development through breast milk.

Xenoestrogens are causing a worldwide epidemic of chronic illnesses and inhibiting natural healing. Reducing our exposure to them, by whatever means we can, is of the utmost importance. Yet every day we eat them, drink them, inhale them, and use them at work, at home, and in the garden. They are present in soil, water, air, and food and are increasing at an alarming rate. The fact is that environmental estrogens are everywhere and can't be completely eliminated. But by maintaining our quorum symbiotic cycles, we boost the health of our innate detoxification systems to protect ourselves. In addition, by refraining from the use of the standard household and personal-care products (many are listed in the Appendix), we will reduce the load of these toxic chemicals in our bodies. Another source of xenoestrogens is solvents that are commonly found in nutritional supplements, cooking oils, laundry detergents, fabric softeners, and cosmetics such as nail polish and nail polish remover. Paints, varnishes, and a wide spectrum of household cleaning products are loaded with solvents. Here are the most common ones to look out for:

- ☐ Alcohols such as methanol
- ☐ Aldehydes such as acetaldehyde
- ☐ Esters, commonly referred to as ethyl acetate
- ☐ Ethers, commonly referred to as ethyl ether
- ☐ Glycols (propylene glycol and ethylene glycol)
- ☐ Halogenated hydrocarbons (carbon tetrachloride and trichloroethylene)
- ☐ Hydrocarbons (hexane, benzene, and cyclohexane)
- ☐ Ketones (acetone and methylethylketone)
- ☐ Nitrohydrocarbons, commonly referred as ethyl nitrate

The effects of the above chemicals are additive and synergistic, which means they are strengthened when used together. Researchers have documented that, in combination, they are thousands of times more toxic than when exposure is to a single chemical.

Personal body-care products also contain xenoestrogens and potential carcinogens. Every time you use a xenoestrogen ingredient on your skin, you alter an ecosystem and suppress your inner physician. Here is a list of the most common ingredients to look out for:

- ☐ **Fragrances**. Many synthetic fragrances (not the real herbal or plant scents) are toxic petrochemicals (petroleum-based chemicals) that enter the body quickly through the skin and accumulate in lipid-rich tissues, such as the brain and nerves, and fat tissues of the body.

- ☐ **Parabens**. These chemicals (propylparaben, methylparaben, butylparaben, or ethylparaben) are cheap synthetic preservatives used to inhibit microbial growth in nearly 99 percent of all cosmetics. However, parabens also promote excess viral activity in your body.

- ☐ **Sodium lauryl sulfate (SLS)**. This is a harsh, caustic detergent used as an engine degreaser and garage floor cleaner. It is commonly found in body soaps, shampoos, and facial cleansers. A potentially dangerous mutagen, SLS is capable of doing serious damage to your DNA.

- ☐ **Mineral oil**. Petroleum-based mineral oil clogs your skin pores and interferes with the ability of your skin cells to eliminate wastes and absorb nutrients.

- ☐ **Imidazolidinyl and diazolidinyl urea**. Commonly used as preservatives, these substances release formaldehyde, which is a powerful xeno-estrogen.

- ☐ **Propylene glycol**. This is a powerful xenoestrogen used as an emulsifying base in creams and lotions to make the skin look smooth. It denatures the skin, blocks meridian energy flow, and is potentially toxic to your liver and kidneys.

- ☐ **Synthetic colors**. Artificial tints are used to color cosmetics and are commonly labeled as FD&C or D&C, followed by a number and a color. (Henna, as a natural plant-based dye, is an exception.)

Chapter Six

Honoring Nature's Recipe for Nourishment

Health and well-being are the result of a dynamic interplay and balance (homeostasis) between human and commensal cells. These cells form strong coalitions and alliances when we are in states of quorum symbiosis with nature. Our ancestors used fermentation as their main method to prepare and store food. This method, replaced by modern technologies, was one of the most important ways for man to stay in homeostasis with nature.

Commensal cells, containing the organizing brilliance of life, are critical for us to excrete toxic chemicals in the environment, and adapt to it. These microscopic living cells are part and parcel of human anatomy. Without them, there is no quorum symbiosis.

Fermented foods embody the wisdom of our ancestors and of nature itself. Marrying the ancient art of fermenting high quality foods to modern methods of quorum symbiosis with quality control, water purification, and drying techniques offers an exciting way to get superior nourishment.

As I stated before, more than two million molecules of the body are made by commensal cells. The complexity of functions of commensal flora is illustrated by the fact that these amazing dynamos contain more than 300,000 different genes, compared to the about 65,000 in the rest of the human body.

The food of our ancestors was more in harmony with nature. It contained only half as much protein, 1/4 as much saturated fat, and 1/10 as much sodium salt as the modern day diet. Most importantly, it contained about 4-5 times more plant fibers, 10 times more antioxidants, 50 times more omega-3 fatty acids, and a billion times more commensal flora and synbiotic nutrients.

It is reasonable to assume that our food habits, which have developed, during the recent few hundred years, could provide explanations to the epidemic in chronic diseases, which has occurred during the last few decades *(Current Opinion in Critical Care 2001; 8; Current Opinions of Clinical Nutrition and Metabolic Care 2001, 4:6; Gut 1998 42).* Stig Bengmark, MD, of Lund University reported all the following beneficial effects from the optimal activity of our commensal cells:

- ☐ Nutrient and antioxidant production
- ☐ Production of growth and coagulation factors
- ☐ Activation of the GALT (gut-associated lymphoid tissue) system, along with modulation and enhancement of many facets of immunological function

- ☐ Destruction of potentially pathogenic microorganisms and their production of toxins (endotoxins)
- ☐ Reduction of mutagenicity to slow down genetic mutations
- ☐ Promotion of growth and regeneration

The modern food industry prevents food deterioration with the use of heat and chemical preservatives. Shelf-life studies look at how temperature, moisture, and packaging methods cause food to deteriorate and go moldy. For example, whole grains that are stored at high humidity and exposed to air will degrade and become moldy within just a few months of harvesting. If the humidity is controlled, and the grains are vacuum-sealed, they will last up to 5 months before going moldy. Considering that grains are harvested in the fall, by the spring they are no longer fit for human consumption. Everything in nature degrades and deteriorates depending on a myriad of conditions. The growths of yeast, mold, spoilage and pathogenic bacteria are to be expected in any whole, raw food.

This being the case, one has to wonder how we can get good nourishment when there's little or no fresh produce available these days. All raw foods that are stored or dried will degrade in nutrient content and palatability until they are void of all nourishment. It is important to remember when discussing the usefulness of various raw foodstuffs and whole grains that there are really two shelf life issues involved.

1. The first shelf life issue concerns the nutrient content of the food,which actually begins to degrade from the moment the food is harvested.

2. The second shelf life issue concerns a food's natural life cycle or the point at which it undergoes undesirable changes in taste, texture, and color and gets moldy or infected with food-borne bacteria.

Food deterioration can be classified into three different groups:

1. **Oxidation of lipids (fats)** – fats oxidize, and when this happens, they become non-polar. Non-polar lipids are bad because they jam up the lipid bilayer of neurons and decrease how efficiently the brain and nervous system work in our bodies. Only fermentation of lipids can keep them polar and, in the original state that nature intended for us to consume them. Yet, the majority of omega fatty acids in vegetable or fish oils are non-polar.

2. **Enzymatic Degradation** – enzymes cause what food scientists call hydrolytic rancidity, which will results in stress on your neurons.

3. **Protein Degratation** - Protein denatures and becomes toxic, clogging up the body's circulatory system and setting the stage for heart disease and stroke later in life.

Our Symbiotic Relationship with Commensal Cells

Humans have lived and flourished with beneficial commensal cells since the dawn of time. Nature uses commensal cells to "prepare" soil-based nutrients for the plants we eat to make them rich in nutrients. In humans, nature depends on the symbiosis or harmony between human and commensal cells to fuel self-healing and spark our levels of innate intelligence.

As ancient farmers observed, and modern organic farmers know, food plants need commensal-type organisms in the soil to form nutrients for optimal plant nourishment. In nature, soil micro-organisms transform rocks, sand, and clay into more complex mineral forms that plants need for nourishment.

Our very intuitive ancestors found out how to harness these micro-organisms to ferment foods and increase their shelf life and nutritive value. Our ancestors were indeed ingenious. With no refrigeration, these fermented foods kept them nourished through the long, hard, winter months. When we consider that virtually every major grain and domesticated animal that exists today was first cultivated or domesticated by our Neolithic ancestors, we should want to honor their tradition of fermented nutrition.

Quorum Nutrition™ refers to a novel process used to ferment foods into precious synbiotic nutrients. If you see the term "quorum-fermented,™ you know it has my seal of approval.

Following nature's recipe, a quorum nutrition ingredient is made with a proprietary blend of commensal cells with raw food prebiotics and supercritical antioxidants (supercritical extraction uses carbon dioxide, the gas that puts bubbles in sparkling water to make super-concentrated herbs that can act to stop fermentation and preserve the raw nutritional cargo of quorum nutrition.

The combination of commensals with appropriate raw prebiotics (food factors and fiber) produces extremely nutrient-dense foods, rich in pre-digested, easy-to-utilize nourishment. These nutients are reduced in weight, from 60,000 daltons in raw whole foods to only 320 daltons, so commensals get nourishment fast, without an expenditure of energy.

The miracle of nature's fermention cycles produces thousands of nutrients that you can't get in any multivitamin or nutritional whole-food product. This instant nourishment results in quourm symbiosis, a process that can only occur when all commensals are nourished and an 8-strain blend is used to colonize the gut.

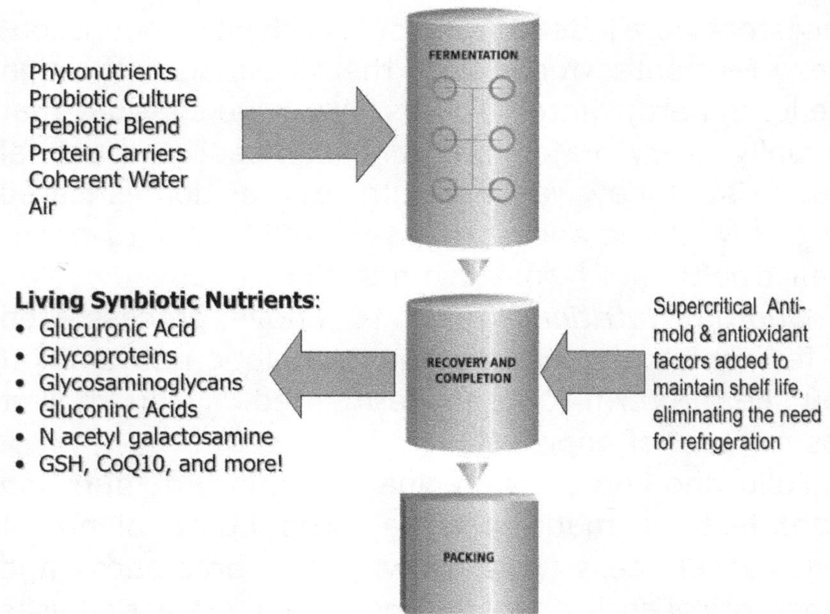

Nature's most precious gift is quorum nutrition. Indeed, every world culture has kept a little of this precious knowledge for our benefit. Foods like natto, miso, and soy sauce in Japan; fermented tofu and soy sauce in China; tempeh in Indonesia; yogurt, chutney, and idly in India; yogurt in the Middle East; fermented grain in Africa; fermented corn in South America; kefir in Russia; sour cream in Europe; pickled vegetables, beer, bread, and wine are used by many people. However, with modern day processing and packaging technologies, most of these products have been heated or pasteurized and, as such, are void of quorum nutrition and beneficial commensal-probiotic organisms.

One has to wonder how probiotic bacteria can survive the high heat of pasteurization. Millions are eating sterile yogurt thinking they are getting some beneficial bacteria. There is simply no way that these probiotics can survive irradiation or pasteurization. Instead, nature rebels and generates high levels of mold or other toxicants. And, indeed our research shows many fermented nutritional products and probiotic supplements are high in toxic mold.

Instead of using pasteurization, heat or processing to preserve the precious nutritional cargo of synbiotic nutrients, *Quorum Nutrition*™ uses nature's wisdom with controlling pH or using supercritical antioxidants to stop unwanted spoilage or mold overgrowth. Supercritical extractions of herbs are nutrient-dense and are powerful solvent-free, concentrates containing the mirror image of nature's intent. They are hundreds of times more concentrated than eating a whole food or herb. Using this advanced technology results in a shelf life of 5 years (no refrigeration required).

Imagine getting raw, living nourishment; no pasteurization or nutrient-destroying chemicals stabilizers, fillers, irradiated ingredients, flowing agents, binders, magnesium stearate (commonly used in the nutritional industry), or GMOs; and, no mold (commonly found in other cultured-fermented and whole-food products).

To maintain the original polarity and vitality of a fresh-picked fruit or vegetable, we use a proprietary *Quantic Harmonization*™ process on all quorum fermented nutrients. Thus, what nature creates in the miracle of fermentation can, for the first time ever, be provided in a daily supplement. I'll be going into greater detail about the impact of these nutrients on overall health later.

Months are spent sourcing and testing raw materials from the cleanest places on earth. Then, all these foods are tested and assayed for authenticity and purity. Sadly, in today's toxic world, only 1 out of 20 organic foods are approved for use in a quorum fermentation matrix.

Like our ancestors that were connected to the healing powers of nature by consuming fermented foods, you can now connect by taking your daily nourishment as nature intended, with thousands of healing compounds and nutrients not found in today's USP vitamins. If you embrace nature's recipes for nourishment, you keep your body's living ecosystem thriving instead of trying to survive life-threatening diseases.

We can reach new limits of strength, resilience, and endurance, despite living in a world so disconnected from nature. And, the repletion of quorum nutrients helps us adapt to the challenges of sustained (or acute) everyday stress. As modern lifestyles grow increasing hectic and toxic, the need for nourishment increases. When quorum nutrients fall below a critical threshold from sustained stress, cell regeneration or innate healing slows down or stops.

Nature fine-tunes the tempo of human life to the biological clocks and amazing rich genetic diversity of commensal flora. Quorum-fermented nutrients have always been a part of life's rhythm. Traditionally valued as a food preservation method, fermentation vastly improves the taste, digestibility, and the nutritive value of foods we eat.

And, since our inner physician relies on neurons to function optimally, quorum nutrition,™ being polar in nature, can get our neurons nourished fast and help them work more efficiently. When this happens, efferent neurons are counterbalanced by afferent neurons, and our body does less damage and more regeneration and repair.

Modern food engineering, including partitioning, modification, and processing, is a far cry from the traditional fermentation methods that yield quorum nutrition. Mass-produced and packaged foods, and nearly all vitamins, are fabricated for us by chemists and engineers to extend shelf life to increase industry profits and decrease our health. These counterfeit foods have no quorum nutrition.

Think about it: Today's food scientists have the nerve to take a farm product and partition it into its individual macronutrient component groups (proteins, carbohydrates and fats). Separating these foods into major groups is <u>not</u> nature's plan. Fractionation into individual components (such as amino acids, isoflavones, fatty acids, vitamins, essential fatty acids, etc.) destroys the original recipe nature had in mind for us to get nourished. It will take scientists another 2-3 decades to identify the commensal cells in our gut and each of the thousands of nutrients made by commensals cells. Indeed, dissolved, heated, flavored, colored, "fortified," "enriched," preserved, molded, extruded, textured, packaged, frozen, vacuum-dried, freeze-dried, tetra-packed, shrink wrapped, boxed, pasteurized, irradiated, and electron-beamed foods are not whole and cannot nourish our inner physician. What a sharp contrast between nature's wisdom of traditional "old fashioned" techniques, such as pressing, pickling, marinating, soaking, smoking, fermenting, drying, salting, and other processes that perverse the full nutritive value of foods!

Dr. Marion Nestle of New York University says, "*If nutrition science seems puzzling, it is because researchers typically examine single nutrients detached from food itself*... and, "*this kind of research is reductive in that it attributes health effects to the consumption of one nutrient or food when it is the overall dietary pattern that really counts most.*"

Modern-day food fails to nourish the body because of widespread irradiation, shelf-life deterioration (food nutrients decrease about 25 percent each day after harvesting) and our inability to digest or extract nourishment from foods. Today's world is no longer in symbiosis with nature as it was hundreds and thousands of years ago. Our physical shape, the quality of our movement, our flexibility, and even our nourishment: These have all been severely compromised. This is why ancient principles of herbal medicine and modern-day synthetic nutritional-nutraceutical formulations with single nutrients as found in your daily multivitamin, or other "milligram-dosed" vitamins and minerals, fall woefully short of nourishing your body.

Beware of Counterfeit Nutrition

Nature's recipes for getting the body in tune are vastly different than the man-made creations of nutritional and pharmaceutical science. Nature employs a *supercontinuum* of reciprocal, harmonic polarities, only found in living, quorum fermented foods, to heal the body. What nature choreographs in elegant and extremely complex molecular artwork can't be duplicated by man. The molecular architecture of quorum nutrition™ allows food to remain alive for a long period of time. Unlike fresh produce, there are no shelf life issues where nutrients decrease dramatically each day after harvesting. Instead, quorum nutrition™ contains whopping doses of thousands of health-promoting nutritional compounds with a long shelf life.

As I stated earlier, I had to invent these quorum technologies, as no one in the industry had figured out how to make non-moldy, unadulterated, and non-heat-damaged quorum nutrition. As my colleagues and I discovered, quorum nutrition was chock-full of stored energy (positive and negative polarities) and nutrients that carried nature's recuperative powers. They pack a powerful knock-out punch to free radicals before these renegade molecules are able to punch holes in our human and commensal cells.

Bear in mind that quorum nutrition is the inner physician's primary source of fuel. That's undeniable. Unlike synthetic nutrition that gives you a short burst of energy, quorum nutrients provide sustained nutrition that the body can use to generate constant energy, build muscle, repair tissues, fight infection, and perform a host of other vital routines.

Since the fundamental causes of nutritional deficiencies are escalating states of environmental toxicity and the daily ingestion of synthetic vitamins and food additives, it's no surprise that people are gaining weight uncontrollably. Let's face it: People perform chemistry in different ways than Mother Nature does. According to Dr. Terrence Collins of the University of Auckland in New Zealand, "...*synthetics are so different from the products of natural chemistry that it is as though they dropped in from an alien world*."Paracelsus, the father of pharmacology, had it right when he stated, "*All that mankind needs for good health and healing is provided by nature*."

The contamination of the essentials of life (water, air, and food) and the quorum symbiotic cycles of nature creates disastrous repercussions, affecting weather, as well as marine, animal, and human life. This contamination has disrupted the functional fabric of nature and our connection to its vast healing polarities. In an Oprah interview, Bobby Kennedy, Jr. stated: "*When we destroy nature, we diminish ourselves and impoverish our children....at our own peril*." Regarding our deteriorating environmental infrastructure, Kennedy feels it is because "...*polluters treat the planet as if it were a business in liquidation and convert our natural resources into cash as quickly as possible.*"

The gravity of these environmental issues has motivated many leading-edge thinkers to reclaim the sense of connectedness that entwines us with nature by only consuming nature's recipes for nourishment. Because the media appeals to prurient interests at the reptilian core of our brains, and all pollutants and synthetic man-made chemicals carry a positive ionic charge, we lose our polarity and disconnect from nature. We crave processed and cooked foods, sex, alcohol, sugar, caffeine, and celebrity gossip. These addictive patterns warp and disrupt our physiology. At the inner physician level, it is analogous to disconnecting the negative terminal on your car battery and expecting your car to run. Just like a battery needs both the positive and negative charges to work, the body requires these reciprocal polarities to function efficiently. Staying healthy from day to day requires that the systems of our bodies maintain the correct polarity states. Circulatory, respiratory, metabolic, endocrine, neurological, lymphatic-immune, and bio-energetic systems operate efficiently and automatically when there is a strong connect to nature's polarities.

When reciprocal polarity is lost, we lose weight and routinely regain it; we vow to eat healthfully and almost always crave addictive foods or drinks. Nearly all degenerative diseases are on the rise, as is cholesterol, blood pressure, blood fats, and blood sugar. Americans are collectively about 5 billion pounds overweight. "*The scourge of body-weight deregulation has become a leading cause of death worldwide*," says Dr. David Cummings of the University of Washington. The biggest wild card in the diet game relates to a loss of digestive capacities and polarities at the cell level. When hepatocytes (liver cells) lose polarity, they fail to produce optimal bile and make the gut uninhabitable for our precious commensal cells. Likewise, when the circulatory system loses polarity, toxins with a positive ionic charge stick to the linings of the vessels, causing cardiovascular disease.

Vitamins that you might be taking can interfere with, slow down, or prevent your inner physician from keep you healthy. These caffeine-like pick-me-ups are addictive. Synthetic vitamins and inorganic minerals can cause insomnia, irritability and nervousness, heart palpitations, and digestive disturbances. They diminish bile production and disrupt the body's healing polarities. When the liver can't produce enough bile, toxins back up into other parts of the body, causing inflammation and degenerative disease.

Infomercials, magazines, the Web, and bulk e-mails promote hundreds of nutritional and herbal products. But is that what your body needs?

A review of human history shows that people in the not-too-distant past were far more robust, more athletic, and leaner than the average American today.

The pharmaceutical and nutritional industries have successfully created the myth that synthetic vitamins and inorganic minerals in precise milligram amounts can be utilized by the cells of the body to enhance health and longevity. Synthetic means "man-made," and nothing synthetic is ever found in nature. All synthetic chemicals disrupt quorum symbiosis and diminish the function of our inner physician. All milligram dosed vitamins have a positive ionic charge that short circuits our inner physician. Thus, beware of products labeled "natural" that are full of U.S. Pharmacopia (USP) synthetic ingredients.

The most powerful stressors and inhibitors of our inner physician are USP synthetic vitamins, inorganic minerals, and toxic body care products. Nobel Prize laureate, Dr. Albert Szent-Georgi, who discovered vitamin C, found that he could never cure scurvy with synthetic ascorbic acid itself. Yet, he reported always curing scurvy with vitamin C found in foods, and concluded that cell utilization requires a food matrix of nature's co-factors. Studies prove that antioxidants and other nutrients derived from foods promote health and are effective in the prevention of cancer and degenerative disease.

Sales for synthetic vitamins and drugs involve a 3500% profit margin, while whole-food quorum fermented supplements are only marked up 20 to 50%. Could industry profits on sales of synthetic vitamins be blinding them to the truth of credible research that says they are harmful and even lethal? Synthetic vitamins are in the majority of today's nutritional products and are found in enriched breads, pasta and flours, and other foodstuffs.

The Scientific & Clinical Evidence against Synthetic & Inorganic Nutrition

Vitamin A was first discovered in 1919. By 1924, pharmacists broke it down and separated it from its natural whole food complex, calling it "purified." By 1931, LaRoche - one of the largest pharmaceutical companies in the world—synthesized vitamin A or made a chemical copy of a fraction of naturally occurring vitamin A, cutting out nature's recipe of other needed components that include *retinols, retinoids, retinal, carotenoids, carotenes, essential fatty acids, Vitamin C, Vitamin E, B vitamins, Vitamin D, enzymes, and mineral-protein matrixes*. Following is a list of some of the USP synthetic vitamins that you will find in 99% of all multivitamin or nutritional products:

Vitamin	Fractionated USP Synthetic Vitamins
Vitamin A	Acetate, Retinal Palmitate, Beta Carotene
Vitamin B1	Thiamine HCI, Thiamine Mononitrate (coal tar derivatives)
Vitamin B3	Niacin
Vitamin B6	Pyridoxine
Vitamin B12	Cyanocobalamin
Vitamin C	Ascorbic Acid, Pycnogenols (from corn sugar/syrup)
Vitamin E	d-Alpha Tocopherol, d1-Alpha Tocopherol, d-Alpha Succinate (from processed food oils – cottonseed, soybean)
Vitamin K	K3 or Menadione
Folate	Folic Acid

Dr. Agnes Faye Morgan (University of California) reported that taking synthetic vitamins is worse than starvation. Animals fed synthetic vitamins had toxic reactions or died quickly of degenerative diseases compared to those fed whole foods (*Journal of Biological Chemistry 1927, 7484; Journal of Dental Research 178, 57; Farmakol,Toksikol. 1979,42; Helv Paediatr Acta 1986, 23; Journal Vet Medical Society 1995, 57:5*).

A study of 29,000 Finnish smokers proved that synthetic vitamins increased death rates significantly enough to stop a 10-year study prematurely (*New England Journal of Medicine 1994,330*). A Harvard study of 22,000 physicians reported no health benefits from synthetic vitamins, while other studies report toxicity and serious side effects (*FDC Reports: The Tan Sheet. 1996;4; Nutritional Toxicology. 1982, 1; American Journal of Clinical Nutrition 1990;52;*

American Journal of Clinical Nutrition 1989;49; Lancet 1996,347; American Journal of Public Health 1995, 85; Arch. Pediatrics 1957, 74; Medical Journal Austrailia,1973,2; Lancet 1993, 342; Lancet 1978, 2).And, synthetic beta carotene blocked antioxidant activity and anti-cancer activity of 50 antioxidants (carotenoids) in the diet *Lancet 1996, 347; American J Clinical Nutrition 1995,62*).

Some brave nutritional scientists even went as far as saying food is the best medicine against cancer, reporting that a diet rich in fresh fruits and vegetables has been shown in 128 out of 156 dietary (epidemiological) studies to be protective against cancer *(Journal of the American Nutraceutical Association,* 2000, 3:3).Jerome Block, M.D., of UCLA Medical Center stated that foods "*...reduce cancer risk, clinical cancer occurrence, and/or interrupt the carcinogenic process in appropriately-defined populations."*

Since studies reveal a progressive depletion of antioxidants in our food, Americans are taking isolated supplements. Yet, as Rita Ellithorpe, MD, stated, "*These antioxidant preparations do not match nature's recipe*." Dr. Ellithorpe's research documented that whole food supplements boosted antioxidant activity 239.7% over isolated single nutrients/antioxidants. She suggested that "*...consumers attempting to derive antioxidant protection with pharmaceutical-like preparations or isolated or combinational antioxidants appear to be failing*."(*Journal of the American Nutraceutical Association 2001,4*) .

Studies reveal that the prolonged consumption of adverse diets (supplements), exposure to water and air pollution and adverse lifestyles have increased cancer rates dramatically. Widespread consumption of refined flour (USP enriched) has been associated with increased cancer induction, accelerated tumor growth, and metastasis (*Nutrition & Cancer* 1983, 4), while the common dietary practice of using margarines and refined oils (high in trans fatty acids) enhances tumor growth and carcinogenesis (*Nutrition & Cancer* 1999,34; *Preventive Medicine* 1990; 9).

In rats, colorectal cancer risks are only reduced with tomatoes rich in lycopene; not with isolated lycopene. These studies clearly show that isolated antioxidants are not protective against our toxic environment *(Japanese Journal of Cancer Research 1998,89*).

Keep in mind that on a daily basis, your body is exposed to environmental toxins. They're in your food, your place of work, and in your home. The human body that is out of sync with nature warehouses toxins in deep-tissue places (bones, connective tissue, and lymph channels), later causing cancer or other life-threatening diseases.

As you have learned, quorum nutrition is the material basis for igniting reciprocal, bi-polar healing energies that activate and enhance detoxification and regenerative functions of your inner physician. But, it is critical that supplemental nutrients are vibrant, quorum fermented so they are in low molecular weights for rapid correction of nutrient deficiencies at the cell level.

These nutrients have the potential to restore digestive, hormonal, and immune functions. They are powerful detoxifiers that disarm pollutants so your inner physician and its neural networks can excrete them out of the body on a daily basis. Without quorum nutrients, the body cannot keep itself clean. Toxins infiltrate and stress the immune system. Immunological warfare is initiated that burns out the immune system, causing it to deposit toxins in the skin, and lymph nodes forming tumors, fibroids, and cysts.

When the immune system is stressed and malnourished, it doesn't have the strength and stamina to conquer microbial foes. Candida, runaway viral infections, and stubborn biofilms accumulate in the body. And, contrary to popular opinion, synthetic vitamins, antimicrobials, anti-yeast, anti-viral botanicals, or cellular resonant therapies make these issues worse. Instead, I teach doctors to put more emphasis on the incredible reciprocity of nature's own cleansing and nutritive agents and the harmonic bi-polar forces that entwine us with nature and weave the web of our existence.

When the hormonal system is malnourished, the effectiveness of the blood-sugar regulating hormone, insulin, declines. We store energy as fat instead of using it in our cells for energy. The excess abdominal fat sparks inflammation that damages the body's organs and glands. The inflammation traps toxins in the body, and we become increasingly toxic.

Chapter Seven

Anti-Quorum versus Pro-Quorum Tactics

As you reflect on the evidence against what the overwhelming majority of people take as a daily multivitamin, you undoubtedly realize that man-made chemicals break up the cellular coalitions and alliances needed for your body's inner physician to execute your physiology. The key to stimulate coalition is to identify "the bad guys" or the common enemies of quorum symbiosis. The following are the most common disruptors of quorum symbiosis:

1. **Synthetic vitamins** and inorganic minerals, nutraceuticals and pharmaceuticals, and hormones that are milligram dosed. Beware of Anti-aging based hormone replacement therapies (HRTs) with progesterone, DHEA, pregnenolone, HGH, or melatonin, as they work to disrupt quorum symbiosis. Even progesterone used as topical sterol diosgenins (wild yam) or synthetic progestins in females can induce leaky blood

vessels, activate inflammation, increase risk of cardiovascular disease, and encourage cancers to spread rapidly and metastasize (*Rheum Dis Clin North Am*, 2005; 31; *Arch Neurol*, 2005; 62; *J Clin Pharmacol*, 2005; 45; *J Clin Endocrinol Metab,* 2005; 90:1181-8; *Fertil Steril*, 2005; 84).

2. **Digestive enzymes** or hydrochloric acid derange the critical ecosystem needed for commensals to survive and enter quorum symbiotic cycles. The side effects of failing to activate your body's own digestive enzymes with commensal cell symbiosis are leaky gut syndrome, ulcers, opportunistic yeast infections, and prolonged gut inflammation.

3. **Herbal and food concentrates** like oregano, neem oil or tea, garlic extracted allicin, bee propolis, and super-concentrated goldenseal root lomatium and colloidal silver may quell a local infection but in the process, they slaughter your commensal cells which are 70-80% of your immune system. Gaining resilience to pathogens requires quorum nourishment.

4. **Man-made chemicals** in body care and household cleaning products or fragrances used to cover up odors (see Appendix for safe non-toxic products).

5. **Addictive substances** like caffeine, alcohol, or sugar. How do you know if you are addicted to vitamins, caffeine, or sugar? Try eliminating them for three days. See how you feel. If you get a headache, feel tired and depressed, or anxious,

you are functioning on stimulation. If your body craves stimulation to function, you're addicted. While in a mode of needing stimulation, innate healing is suppressed, and stress cannot be compensated for without damage to your body.

6. **Mold or mycotoxins** - Today, mold or mycotoxins may present the greatest health threat. While much attention lately has been given to the dangers associated with stealth infections, it's surprising that the dramatically increasing health threat of mycotoxins—secondary metabolites produced by many species of fungi—is still being largely overlooked. Mycotoxins are naturally occurring chemicals that are produced by fungi growing on feed, food, or grain. These fungal metabolites are highly toxic and highly suppressive of the immune system. Unfortunately, they are increasingly present in processed foods, especially in peanuts, tree nuts, beans, apples, grains, and cereals. In 1985, The Food and Agriculture Association estimated that 25 percent of the world's food crops are contaminated with mycotoxins, and the number is probably much higher today. According to the World Health Organization, mycotoxicosis can cause Alzheimer's disease, multiple sclerosis, atherosclerosis, and cancer. In addition, many studies have documented that mycotoxicosis is a causative factor in multiple chemical sensitivity syndrome, as well as in respiratory and neurological disorders. Iris R. Bell, M.D., of the University of Arizona Health and Sciences Center, showed abnormal brain-wave activity among patients exposed to mycotoxins. Bell and her

fellow researchers documented that mycotoxins have a direct biological role in initiating and/or perpetuating nervous system–related illness. The connection between the environment and health, according to A. V. Constantini, M.D., from the University of California, School of Medicine, San Francisco, is not that major diseases are caused by the consumption of specific foods, but that they are caused by the fungi and mycotoxins present in the food chain. Try to avoid eating moldy food, and clean up mold in your basement.

Pro-Quorum Tactics: A Cooperative Team Game

Nature has declared life to be a team game. Your own body is testimony—from the teamwork of your genes to the quorum cell talk that forms your body. Life is governed by cooperation between cells. In the case of commensal cells, some of them die to benefit our human cells. Like bees that die in the act of stinging, the salmon that die after they spawn, or the worker ants that help the queen breed and secure the next generation, our commensal cells work hard to keep us alive and well nourished.

Pro-quorum tactics in nature use cooperation to ensure the integration and coordination of our body's components. These cooperative efforts are mandatory to long-lasting health. My research shows that we can initiate this cycle if we re-colonize the gut with the **8-Strain Commensal Probiotic Colonizer** that empowers the afferent regenerating neurons and the digestive and detoxification process. But we have to feed this 8-strain blend with quorum nutrition for them to proliferate and differentiate into the full spectrum of commensal cells originally intended by nature before antibiotics.

A quorum of "Commensal cells mediate intestinal epithelial cell protective responses, including enhancing barrier function, proliferation, differentiation and cell survival."

Current Opinions Gastroenterology 20:2004 Valderbilt University School of Medicine

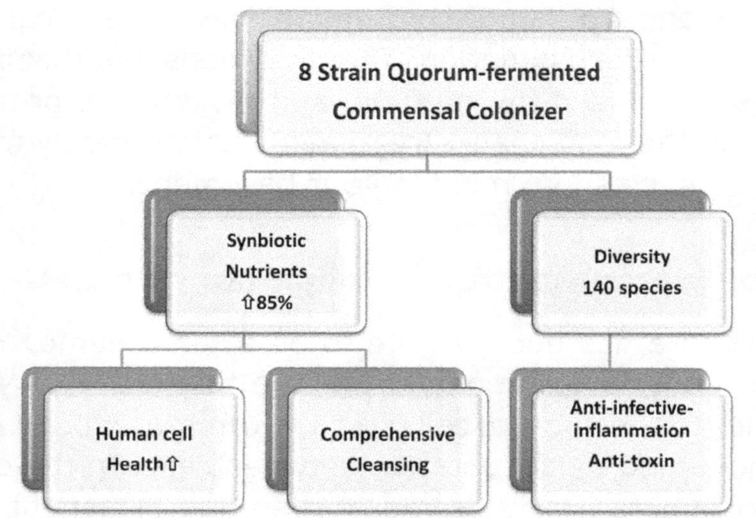

Nature's power source is from raw quorum fermented nourishment. A ton of research, especially in medical journals, confirms what nature has shown us regarding anti-quorum tactics that disrupt and derange our gut ecosystem. According to Stig Bengmark, MD, PhD, of the University College of London, one of the world's top authorities on probiotics "*...combining probiotics and prebiotics into 'synbiotics' will further enhance immune-supportive effects...A deranged gut environment, deranged flora and mucosal lining, has during the last thirty years increasingly been recognized as a source of allergic and autoimmune diseases, but also of acute and chronic infections*."

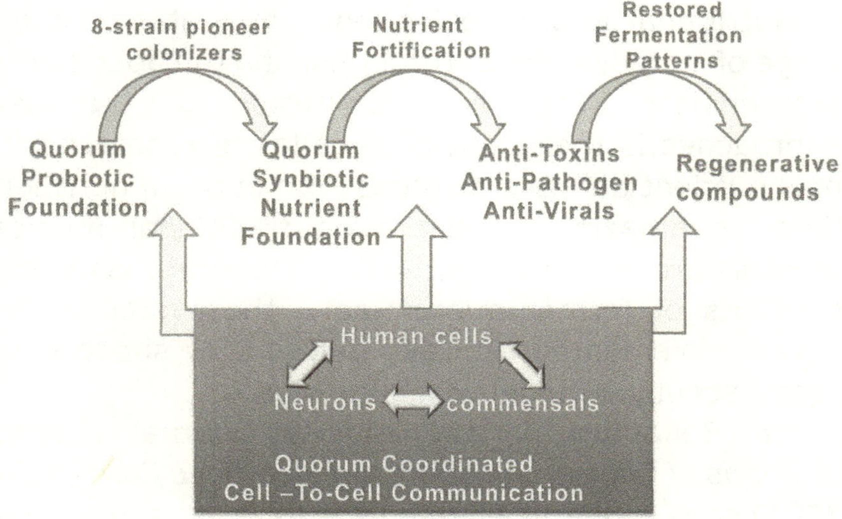

Leading authorities on the gastrointestinal tract are telling us that we need to go back to the diet of our ancestors to be healthy. Recall that this diet consisted of some 500 plants (we eat less than 50 plants), as well as raw, fermented foods (we conserve, dry, and cook our food, processes known to destroy many sensitive nutrients and antioxidants). When commensal cells are nourished this way, they digest our foods and transform them into living quorum nutrition, which ignites powerful regenerative cycles throughout the body. These incredible cells continue to feed and nourish us even when we are not eating!

Sadly, the majority of Americans have very little understanding of the necessity, function, and benefits of beneficial commensal cells and their quorum symbiotic cycles in nature. Instead, processed and refined foods and mega-potency multivitamins are used indiscriminately. Unknowingly, these consumers suppress inner physician dynamics by triggering exaggerated stress responses that can cut their lives short.

Nutritional and herbal science now stand on the verge of incredible breakthroughs, and the possibilities are endless. The most promising of the new approaches is the result of decades of research into new nutrient delivery methods with quorum nutrition. What is amazing is that even traditional medical journals are in agreement with the quorum symbiotic concepts of fermentation. Sadly, these studies are never implemented clincially unless they support the pharmaceutical model of health care.

Dr. Bengmark in the medical journal *Current Opinions of Clinical Nutrition and Metabolic Care* (4(6): 2001) says it clearly when he states, "...*It is likely that several hundred, thousand, if not millions, of synbiotic compounds are released by microbial fermentation, and are subsequently absorbed. Among these are various short chain and other fatty acids, amino acids, peptides, polyamins, carbohydrates, vitamins, and numerous antioxidants and phytosterols. More than 4000 plant flavonoids have been identified, 600 carotenoids, and some have the antioxidant potential 10 times as strong as vitamins C and E.*"

Quourm Nutrition is synbiotic in nature and id made up of living nutrients organized into very complex patterns that are immeasurably effective at keeping our bodies clean and healthy. At very small doses, they have superior potency and bioactivity compared with megavitamins that work adversely to accelerate stress reactions. Low-dosed *Quorum Nutrition*™ gives us living nutrition that mimics the gentle and subtle rhythms of the body's inner physician, keeping us balanced and super-resilient to stress.

When deficient, insulin, the hormone that regulates blood sugar, can cause cancer (*Neoplasia, 2009:11; FEMS Microbiology, 2002, 217*) and heart disease (*J Am Diet Assoc, 2009, 109*). *Quorum Nutrition*™ provides complex carbohydrate molecules called "glycans" that can suppress tumor growth (*Proc Natl Acad Sci USA, 2009*) and stabilze insulin so we do not store sugar as fat in the abdomen.

Researchers have known for a long time that obesity may be the cause of 25-30 percent of cancer (*IRAC Handbooks of Cancer Prevention, Lyon France IRAC Press 2002*). Another study found that 14 percent of all deaths from cancer in men, and 20 percent of all deaths of those in women, are due to obesity, with a strong link to the hormone insulin (*New England Journal of Medicine 2003; 348*).

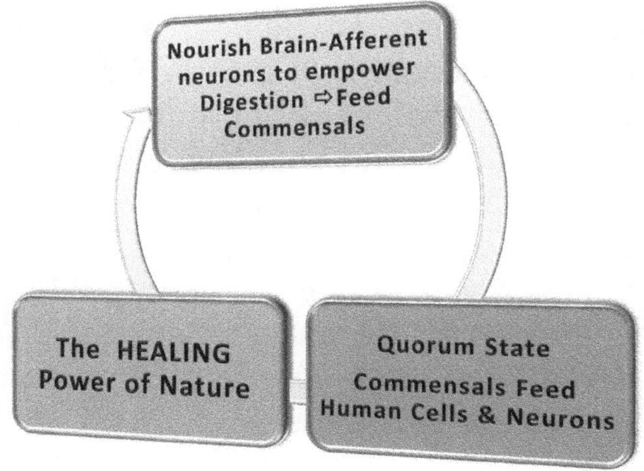

A leading oncologist, Dr. D. Barry Boyd, at the Greenwich Hospital in Connecticut suggested that insulin might speed the cellular growth of cancers. Another researcher found that refined carbohydrates—found in white flour and processed foods—increased cancer risk by disurpting the function of insulin (*Cancer Epidemiol Biomarkers Prev 2004, 130*).

The ingestion of living, raw, food-based nourishment, bursting with an array of vitamins, enzymes, proteins, minerals, and energy, can turn on the full power of nature by re-tuning our inner physician to the organzing brilliance of life itself.

The single best advice for living a long, healthy life is to choose your nutrition well. You'll learn more about nourishing, easy-to-digest and anti-oxidant rich supplements, praised by scientists and used for thousands of years in different world cultures, later in this book.

As you discover the potential you have to awaken your inner physician, you'll learn how to create optimal health. You'll enhance your intuition and your health in concert with these subtle and cooperative rhythms of life in the pro-quorum cycles of cells.

Becoming aware that you need a solution to today's toxic and fast-paced world, stirs up hopes that, with the right kind of nourishment and information, you'll get better. Fears about the unknown origins of an illness disappear, and with the right information, feelings of guilt dissipate. Once your wholeness and harmony are restored, your body will be less susceptible to assaults, whether chemical, infectious, or emotional. You'll have natural antidotes to compensate for the wounds and bruises inflicted by stress, and they will make you more resilient, more confident, and stronger.

Quorum Nutrition™ is your first line of defense against pollution and stress. It is the compensation you need to stop stress from eroding your body and zapping your energy. Having a nourished and clean body determines your performance efficiency or overall resilience to damaging effects of stress. When your performance efficiency is optimal, your body has the potential to do more than is usually necessary to ward off the negative effects of stress or toxicity. One more point: This also means that your best health insurance is to (1) avoid anti-quorum stressors that overpower or diminish your inner physician, and (2) optimize your nourishment to de-stress your body.

Chapter Eight

Embracing Quorum Nutrition in Everyday Life

"A human being is part of a whole, the universe. Our task must be to free ourselves from the delusion of separateness, to embrace all living creatures and the whole of nature."

Albert Einstein, Physicist, Genius, Professor

Approaching and regarding life with a reverence to nature allows us to see our lives differently. As you acquire a sense of reverence for the wisdom of nature, you will look at the food and supplements you eat and the water you drink much differently.

One of the best ways to cultivate the ability to see through myths and health frauds is to recognize them for what they are. Always be aware of how little you know and how much your inner physician knows of the countless facets of your daily life.

You are about to make a substantial change in your life that may seem difficult at first. It's going to get easier because most people feel the rewards right away. You'll sleep better, feel better, and have more energy.

Your body is teaming with living cells. Living cells need to derive, store, and use energy from living nutrients. Now that you have discovered the facts about nature's recipe for nourishment, it will be easier to make the necessary lifestyle changes to get healthy.

By choosing your supplements wisely, you'll be amazed how stress will bounce off you and how you will never have to reach for sugar or caffeine to stay functional at the end of the day. The quorum nutrition™ designation on the label of your nutritional product, will help you choose the right supplement out of the thousands of products available today.

Living cells need nutrients that vibrate with polarity and energy. Our neurons need these positive and negative polarities to empower our inner physician. It's a little like charging the battery in your cell phone so you have many hours of cell phone usage.

Why keep our neurons charged? From the level of cut fingers and into the mental realm, where emotional stress overload requires compensation, the brain works its healing miracles to keep us in good health. Unparalleled by ordinary supplements, *Quorum Nutrition*™ allows for rapid and unmatched nutrient uptake so you can protect yourself from pollutants and stress. In the strictest sense *Quorum Nutrition*™ is a food. Foods are extremely complex mixtures of bioactive constituents and living energies (life principle) that work together to promote health in the human body. As I previously explained, there is a vast difference between modern diet and the diet of our ancestors who ate 5 times more fiber than we do from over 500 plants. Americans eat from less than 50 plants and rarely eat them raw and fermented, as did our ancestors. Instead, food is conserved, dried, and cooked with a dramatic reduction in fermented, quorum nutrients. Thus, daily supplementation with pure, raw fermented food concentrates derived from the cleanest places on earth, is critical to restore our quorum symbiotic cycles.

Quorum nutrients are transported to the cells with carrier molecules so that they are passively diffused into cells with no energy expended. Since mold is high in fermented foods and supplements, I developed a unique quorum process that would generate copius amounts of pre-digested nutrients for instant cellular utiliztion.**

University studies document that cell utilization for fermented antioxidants is 20.8 % to 238% greater than for isolated USP synthetic vitamins. Using Nobel Laureate research on how proteins carry information like "zip codes" to our cells, a fermentation process allowed proteins to be joined to their inherent counterparts (minerals, enzymes, etc.) via nature's wisdom for rapid uptake by our cells (*Blood Coagulation & Fibrinolysis, 2004,* 15:8; *British Journal of Clinical Phytomedicine 2002,* 11:4;. *Journal of Medicinal Foods, 2001, 4:4; Agricultural and Food Chemistry, 1998,46:4; Nutritional Research 1998, 18:4; Nutritional Biochemistry, 1996, 7; Irish Journal of Medical Science, 1993, 162:7; European Journal of Cancer Prevention, 1993, 2:1;.Gut, 1993, 34; Medical Science. Research, 1992, 20; Nutrition Research, 1992, 12;. American Journal of Clinical Nutrition 1988, 48:3; Nutrition Reports International, 1987, 36:3; Nutrition Reports International, 1985, 32:1;, Nutrition*

Reports International, 1984, 30:4; Nutrition Reports International, 1983, 27:4; Science. 1997, 278; Townsend Letter for Doctors, Dec 2000; Journal of Applied Nutrition 1988. 40:2).

There are doubtless myriad therapeutic applications to applying Quorum Nutrition™ in the clinical world. Already, surgeons like Dr. Bengmark are applying these principles, noting vast improvements in post-surgical recovery time and healing.

Doctors stuck on the use of synthetic USP vitamins or amino acids may quell symptoms in the short term, but long-term jamming of quourm signals eventually leads to more serious ailements and the use of more and more pharmacetical medicines. Man-made chemicals disrupt nature in our bodies.Instead, quorum fermented proteins, rigged by nature, can carry huge toxic payloads of toxins out of the body. ** Fast clearance by binding to toxins, pathogens, or aberrant signals—offered by nature—is in harmony with Hippocrates's famous adage, "*Let food be thy medicine and medicine be thy food*."

Seven Steps to Engaging Your Inner Physician

When it comes to following nature's wisdom, fad diets are a recipe for disaster. Quick-fix supplements or diets will inevitably leave you frustrated and discouraged and send you back to sitting on the couch depressed. In order to achieve long-term success, you must approach diet with respect for nature's wisdom.

Your body: It's yours and yours alone, yours to care for, and yours to nurture and nourish. A key element missing from diets is an understanding of why we eat too much. Our cells are starving for proper nourishment!

We have lost contact with nature and fail to digest, assimilate, and nurture our bodies properly. These facts make it almost impossible to get nourishment from supermarket foods where critical nutrients are lost to processing, cooking, irradiation, pasteurization, and sterilization techniques.

In the context of an overworked, overtired, and overextended lifesytle, changing your diet may be difficult. Try to make it fun. Look forward to feeling and looking better. Find foods that you like, and incorporate them into your daily diet. There's no way around the fact that diet, exercise, and the right kind of food concentrate are essential for a healthier, longer life.

Outsmarting the fat cells of your body means that you have to correct the underlying deficiencies that slow you down metabolically. You are only storing fat because you are in starvation states due to maldigestion.

Throw away your calorie counters and your food scales and follow the seven steps below to get your metabolism functioning at peak efficiency.

The following seven steps are aimed at helping you treat your body's inner physician with respect and restore quorum symbiosis:

1. **Don't buy into quick-fix supplements, diet plans, diet pills, or addictive caffeine or vitamin-containing products**. Diet pills, green tea and other teas, coffee, and synthetic vitamins all stimulate a false sense of energy that only makes fat cells fatter. Avoid supplements marketed as "fat-burners" or "thermogenic," which deprive your body of nutrients and rev up

its fat-storage machine. Don't look for gimmicky liquid meal replacements or energy boosting beverages that claim to shift your metabolism into high gear. Make sure the supplements you are taking are not working against your overall metabolism or killing off your commensal cells. Following nature's recipe and wisdom is always the safe and wise course when choosing a source of nourishment to jump-start your metabolism.

2. **Consume Quorum Nutrition daily along with the 8-strain commensal cell colonizer** (see sources in **Appendix A** and advanced information in **Appendix C**) for nourishment far above the quality of any organic food or supplement you can buy in your local health food store.

3. **Eat raw, fresh, whole foods and healthy fats daily.** Nearly all people who are overweight or sick can't digest their foods efficiently. Getting ample fiber from raw carrots, celery and other vegetables is critical to establising and ecological niche in your gut for commensal cells. When you eat raw, fresh produce, you minimize toxic maldigestion reactions and provide a safe haven for beneficial commensal-probiotic cells to proliferate in your gut. You want this army of cells on your side to keep the intestines free of infection and toxins and to digest your food more effieicnty into micronutrients. Evidently with the help of the Quorum Nutrition and the 8-strain commensal cell colonizer, your commensals will find a permanent residence in your gut. Remember, they are the strong arm of immune defenses

and manufacture nutrients to keep your body superhealing modes.Raw sesame butter and raw shredded coconut are the ultimate best fats for commensal cells. Make vegetables, whole-grains, and legumes your main meal. Let fish, fowl, or lean meats serve as condiments. Healthy fats are also found in cold-pressed extra virgin olive oil and raw fresh nuts. Unlike animal fats, these fats do not raise blood cholesterol nor cause you to gain weight. Simplify your diet.

4. **Drink at least 3-4 glasses of purified water daily**. When the need to eat strikes you, you may actually be thirsty. The hunger signal is often confused with the signal for thirst in your brain. For superior hydration add at least one or more ounces of *HydraWater*™ as a catalyst to your purified drinking water each day.** (see **Appendix A** for more information)

5. **Exercise daily.** Walk fast while maintaining good posture. Find a walking buddy, and commit to meeting them regularly. Ask your doctor where to start with walking. Most start with 15 minutes a day and add 5 minutes every week until they are walking one full hour a day. After a few weeks of exercising, you'll have more energy during the day, sleep better at night, and start noticing some positive physical changes in your body. Stay flexible by stretching your muscles daily before and after walking. Keep in mind that one hour of fast postural walking can reset your body's "set point" so you quickly reach your ideal and healthy waistline. While exercising only, always wear the

Quantum E Protector to clear out any cell phone or modern day frequencies that can disrupt the function of your inner physician. (see **Appendix B** for details on posture and exercise)

6. **Minimize stress** by avoiding stimulatory substances and by keeping a positive attitude about life. When it comes to your health, it is most important to make sleep and relaxation a priority. Don't eat close to bedtime! A full stomach interferes with sleep patterns. Breathe deeply in response to a stressful situation. Stress is inevitable! Every tense situation, or even a memory of one, causes a change in breathing, and it's best to take a deep breath. Don't deny or repress your feelings. Maintain emotional balance by sharing your feelings with family and friends. Don't hold negative feelings in, as they will continue to perpetuate harmful cycles of stress.

7. **Lower your exposures** to environmental pollutants, household chemicals, pesticides, toxic fragrances found in body care products (loaded with fat-storing xenoestrogens), irradiated foods, and genetically modified (GMO) foods. Food irradiation, which ostensibly kills anthrax and food-borne pathogens and gives food a longer shelf life, has a decided anti-quorum downside, one that negates any so-called benefits of the process. See **Appendix A** for safe foods and products.

No Quick-Fix or Fast Solutions

If you're ready to apply nature's quourm healing secrets, you have to realize that it will take time to cleanse and nourish your body after you've become toxic and nutrient-depleted. But, fixing the internal biology of your cells has tremendous potential for your long-range health.

When we have lost many commensal cells it takes time to restore quourm symbiosis and the full function of your inner physician. This proces does not happen overnight. It takes years and decades of poor eating and undernutriture before you end up in dysbiosis and with excessive inflammation and disease.

As you begin to cultivate and nurture your inner physician and its cellular counterparts, your body will naturally want to throw off toxins. This means that you do not have to fast on a water or juice diet or go on caloric-counting diets to attain health or lose inches off your waistine .

The body, when healthy, burns, rather than stores, excess calories in a process called metabolism (also referred to as thermogenesis). Nature's furnace for fat-burning is dependent on quorum nutrition. Focusing exclusively on diet to control weight or get healthy fails if your cells are not fueled by proper nourishnent. So far, we have looked at the body's own restorative and rejuvenative powers and their diminishment from being disconected with nature.

Remember, throughout the pages of this book, I have stated that optimal health requires proper nourishment and cleansing of our cells. Doesn't it make sense that, if we do this, we can be disease-free, injury-free, toxin-free, and highly energized in a state of super-metabolism?

These lifestyle changes will cleanse your body of harmful toxins and carcinogens that can alter your brain chemistry, causing fuzzy thinking or changes in your moods, thoughts, feelings, and behavior. As a result, you should begin to notice positive changes in your mental attitude. In addition, staying away from stimulatory supplements or foods can help to stabilize your blood sugar and shift you from a negative to a postive emotional state. For example, people who suffer depression due to toxicity often crave foods that can lead to obesity. Or their depression can cause them to engage in a high-anxiety lifestyle in order to avoid exploring their feelings.

Enjoying life is about feeling energetic enough to take part in activities and relationships that are important to you—that motivate you to get up every morning. How you view life and the satisfaction you get from life have a major influence on your health.

With the information you've learned in this book, you have the potential to enhance your body's innate healing energies and become more resilient to stress. Not everyone responds in exactly the same way to the pressures of life. Some of us are able to rise above crises, while others are left devastated. Much of our ability to cope during, and rebound after, stressful life passages has to do with our overall state of health. Some clues to your own stress level may be found by reviewing habitual behaviors, as well as the messages your body is sending you.

If you're over 50 years of age, you likely won't see changes overnight. And, you may need medical help to normalize your hormones and metabolism. After a period of 4 to 6 weeks, you should notice some positive health changes and experience permanent weight losses. There is no miracle cure for many of today's ills or stubborn belly fat—but you can help to fuel your metabolism with suggestions offered in this book.

Don't delay! Set your goals, and stop wasting time and money on diet fads and quick-fix approaches. Diseases are reaching global epidemic proportions. Think about it. The top three causes of death in this country have been scientifically linked to nutritional deficiencies, and therefore, are preventable. *Quorum Nutrition*™ gives you the most advanced way to help nourish your innate immune system and its commensal cell disease fighters against the nation's top killers.

A quarter of the population has heart disease. As much as 40 percent of those who have a heart attack have no warning symptoms at all. And here's the shocker: For half of those who have a heart attack, the first symptom is death.

But that's not the way it has to be. Good nutrition can prevent heart attacks. Citing 749 scientific studies in a 2002 *JANA* article, Mark C. Houston, MD, of the Vanderbilt University School of Medicine, concluded that optimal nutrition can prevent, delay the onset of, reduce the severity of, treat, and control cardiovascular disease.

Heart attacks usually begin with an infection and plaque in the arteries. Like gunk inside an old pipe, plaque keeps building up. Then, one day, often with no warning, the plaque ruptures, clots, and stops the flow of blood to the heart, causing a heart attack.

Quorum Nutrition™ has the potential to energize you and give you a leaner, stronger body.** Remember, uncompensated stress is our enemy. And experts agree that malnourishment of our stress defense mechanisms causes the body to store fat. In turn, stored abdominal fat has been scientifically proven to induce inflammation and pain. No wonder millions of Americans are dependent on anti-inflammatory or pain-relieving drugs!

Inflammation is the driving force behind nearly all degenerative disease affecting our bones, joints, arteries, brain and immune system. Drugs are failing us. Dr. David Graham, a reviewer for the FDA, said, "*Americans are virtually defenseless against unsafe drugs, and called regulators incapable of protecting America against another Vioxx.*" The steady drumbeat of bad news about the side effects of drugs has left many Americans confused over how to treat pain.

Nurturing and nourishing your body's inner physician makes perfect sense. You are hardwired for health, not disease. Allow your body to undergo reorganization as you tap into the greatest healing force known to man.

The Final Word

In conclusion, before you start this program, I strongly advise that you tell your doctor what you're taking. Ask your doctor before implementing this program, and get his approval. If you have been medically diagnosed with a disease, follow the advice of your doctor, and don't endanger your life by stopping prescribed medications. Opt for complimentary or alternative medicine physicians when there's no sudden or severe, life-threatening health problem.

Continue to rely on your physician to diagnose your health issues.

A word of caution: My breakthrough research on quorum nutriture is quickly becoming accepted by many in the field of alternative medicine. Unfortunately, many companies have scrambled to come up with cheap, knock-off products to copy my unique discoveries. Using buzz words like "nano" "cultured" "fermented" "cell resonant" and or "Quantum Nutrition" they try to confuse the public into thinking they are getting the same type of nutrition I discovered. Beware! Look for the designation Quorum Nutrition™ on the label,as the vast majority of other so-called fermented or cultured supplements do not use quorum methods and may be high in mold and toxic substances that, in my professional opinion, are unsafe and dangerous to health. When misused, fermented foods can undermine the faith of a public primed to expect health benefits from what they are reading in this book.

The thin foundation of science that underlies the nutraceutical and nutritional industry has fooled many consumers. The claims of many web-based purveyors of so-called natural health products provide a cautionary tale of how commerce and making money race ahead of science. The brilliance of nature and the science of food has given us everything we need. It is pretentious and vain to think man-made vitamins and nutraceuticals are better than whole and fermented foods.

If you are stimulated by the ideas in this book, the Appendix offers further ideas and information that will help you begin the practice of awakening the full potential of your body's inner physician. For those of you who are ill, for those who are well, and for those who practice naturopathic medicine or alternative medicine, the message is always the same: Nourish and detoxify all the cells in your body to restore resilience to stress, and tap into the full powers of your inner physician.

Appendix-A: Resources for Products and Further Information

This listing offers addresses and phone numbers and a brief note about product lines. Supplements should be taken under the direction of an alternative health-care practitioner, preferably one certified in Quantum Medicine. Some of these products require a physician's prescription, while others can be purchased in your local natural food store, by mail, or over the Internet. Addresses and phone numbers are subject to change.

HEALTH PRODUCTS, AND RELATED PRODUCT LINES

QuantaFoods, LLC

Fax 386-663-9075 www.quantafoods.com
For laypersons - The authentic and exclusive source for *Quorum Nutrition*™ or quorum-fermented™ nourishment and the **8-Strain Commensal-Probiotic Colonizer** called *Quorabiotics* ™. Also source for the *Quanta e-Protector* to protect against harmful electropollution is also available at the website.
For Doctors If you're a doctor or licensed practitioner, QuantaFoods, LLC also has clinical products to deal with the chronic effects of advanced dysbiosis.

Quorum Nutrition, LLC

3905 E Russel Rd, Suite E
Las Vegas, NV 89120Fax 702-924-2526 www.quorumnutrition.com

For laypersons - Education and personal e-mail consultations by Dr. Paul Yanick

American Academy of Quantum Medicine (AAQM) – www.aaqm.org

Las Vegas, NV This Nevada–based non-profit organization, founded by Dr. Yanick is a professional, accredited research and educational non-profit organization that fulfills an urgent need for clinicians who want up-to-date natural solutions in a practice-focused, easy-to-implement format so they can fully capitalize on practice building, based on results. On this website, I have made FREE publications, videos and lectures available free to the general public. The AAQM's mission is to research and investigate multidisciplinary diagnostic and therapeutic protocols and apply them to current health care practices in an effort to advance health care and minimize human suffering. The AAQM is funded primarily by tax deductible, charitable contributions and accepts gifts in a variety of ways that most befit the donor. Application for Board Certification by American Naturopathic Medical Certification Board (ANMCB) in Quorum Nutrition™ or Quantum Medicine™ is on this website as is many free publications and a FREE video lecture by Dr. Yanick.

HolisticDentalHealth.com

541 N. Palmetto Ave., #101
Sanford, FL 32771 407-322-6143

This holistic dental practice is a father-and-son practice that began in Sanford, FL in 1969 with a mission that continues yet today – to provide the very best dental care based upon the latest research. Dr. George Edwardsd and Dr. David Edwards us non-toxic dentisry along with advanced techniques and sophisticated eqquipment.

Eden Foods Inc products

Yellow Organic Popcorn - A certified organic, carefully selected variety of yellow popcorn with superior popping qualities - large, fluffy tender morsels, bursting with pure corn flavor and no mold! One of the only non-moldy popcorns! It has a great effect at scrubbing mold out of the intestines and should be eaten at least twice a week for best results.

Artichoke Ribbons - A light green ribbon pasta made of organic USA golden amber durum flour blended with organic Jerusalem artichoke powder, organic spinach powder, and purified water. Unlike no other pasta made the Eden Organic Pasta Company, this pasta has a long shelf life and never seems to go moldy over time. Delicious mild flavor, delightful texture and quick cooking.

Mugwort Noodles - Rich, nourishing soba noodles traditionally crafted with buckwheat flour, whole wheat flour, a bit of fine sea salt, and the mild herb mugwort leaf which has a powerful anti-mold effect on the other grains. Mugwort gives this soba its deep green color and delicious flavor. A traditional Japanese, bracing healthy food. A source of protein and iron.

Mung Bean Pasta - Thin, translucent noodles made in the traditional manner from 100 percent mung bean starch and water. Cooks in 3 to 4 minutes. One of the most popular noodles in the Asia, and most frequently the one used in sweet and sour, and hot and sour soups. Their slippery, chewy texture adds delight to soups and salads. Made without salt, wheat and gluten free.

Organic Tomato Sauce - A hearty sauce made from organically grown vine ripened Roma tomatoes, cooked within hours of harvest with organic extra virgin olive oil, a bit of our finest sea salt and a mouth watering blend of organic herbs and spices. An excellent source of dietary fiber and antioxidant vitamins A and C, and a good source of iron. Each serving offers 14.4 mg of the antioxidant lycopene. Mold-free!

Brown Rice Vinegar - Made by a patient 1,000 year old method. Organic brown rice, koji, and spring water are blended in clay crocks and aged 6 to 8 months. Its sweet essence enhances almost any food and is essential to making sushi. 4.5 percent acidity. Protective amber glass bottled with a functional dispenser cap and it's refillable. One of the few non-moldy vinegars available anywhere!

Garbanzo Beans - USA small farm organically grown Garbanzo Beans, soaked overnight and expertly cooked at Eden's certified organic and circle k kosher cannery. Also known as chickpeas, these plump round golden beans are perfect for salad and essential for hummus. Rich in dietary fiber and folate (vitamin B9), and a good source of protein. Low fat and very low sodium with no salt added. Packed in bisphenol-A free cans. Unlike other canned beans, these had the longest shelf-life and stayed mold-free for almost a year.

Oat Flakes - Family farm organically grown in Saskatchewan, Canada. The whole grain is gas fire roasted and rolled into flakes. Ideal for quick hot cereal and more. Cooks in 3 minutes. Heart Healthy* 100 percent whole grain. Rich in dietary fiber and thiamin (B1), and a good source of protein, iron, and magnesium. Packaged in a protective reclosable pouch.

LENTILS - Lentils cooked in a delicious, perfectly seasoned sauce. USA family farm organically grown, carefully chosen and expertly cooked at Eden's certified organic, kosher cannery. Convenient, versatile and ready to eat. Serve as is with grain, or simply add a can of EDEN Tomatoes and your favorite vegetables for soup. Mold and fat free and a good source of dietary fiber and protein.

RiceTec, Inc., Alvin, TX – Tel: 1-800-232-RICE -The organic JASMATI is the most popular because of its unsurpassed flavor and aroma and ability to resist mold. While RiceSelect offers a wide variety of delicious rice and other dishes all grown and packaged in America, get the organic JASMATI.

Ineeka, Chicago IL USA - www.ineka.com

Himalayan Green Tea - One of the finest quality, non-irradiated (in irradiation-proof metal container) organic, non-moldy green teas in the world! Unlike other teas, green tea helps to cleanse mycotoxins out of the body.

Hurley Farms Organic Extra Virgin Olive Oil is rich in texture and lively in color. This is a healthy alternative to butter or margarine and one of the few organic olive oils that is free of non-polar lipids and oxidized fatty acids . www.NapaValleyProducts.com

Gold Mine Natural Foods™ 7805 Arjons Drive San Diego, CA 92126-4368 Tel: 858-537-9830 Fax: 858-695-0811 www.goldminenaturalfood.com

No other soy sauce in anywhere is mold-free and comes close to this 100% organic, non-GMO Nama Shoyu in flavor or quality. It's the only soy sauce that's aged for four years in cedar kegs by a unique double-brew process, so it can be made with less salt naturally, while still retaining its full-bodied flavor and delicate bouquet. Nama Shoyu is also unpasteurized (read: the only raw soy sauce available in North America); it's full of health-giving live enzymes and beneficial organisms like lactobacillus. Ingredients: Organic whole soybeans, mountain spring water, organic whole wheat, sea salt.

Eddie's Pasta is a complete line of organic pasta. The vegetable pasta not only has exceptional taste with no added color – it gets its vibrant color from vegetable powder. Eddie's whole wheat pasta is made with 100% organic durum whole wheat and stays mold-free for longer periods of time compared to other pasta on the marketplace.

HEALTH PRODUCTS, AND RELATED PRODUCT LINES

Alta Dena Industry, CA 917441-800-milk-123 (1-800-645-5123)
Manufacturer of raw (non-pasteurized) cheddar goat cheese and kefir. No colors or preservatives added. Certified rBST-free. Order direct to avoid irradiation via shipping.

Environmental Health Coalition of Western Massachusetts (EHCWM) P.O. Box 614 Leverett, MA 01054
A grassroots organization created to help educate the public about environmental health issues and meet the needs of chemically injured people.
Lumiram Electric Corp. White Plains, NY 10606
www.lumiram.com Manufacturer and distributor of Chromalux full spectrum light bulbs.
Micro Essential Labs P.O. Box 100824 Brooklyn, NY 11210
Customer Service: 718-692-3618 Manufacturers and suppliers of pH tape for home testing.

Common Toxic Commercial Product Ingredients and Potential Health Effects

Taking in food that is toxic to the system is one way, as we have seen, to abuse our bodies and lower our quantum energy fields. However, we can also bring toxins into the body through substances we breathe in and others we use in our daily lives, things like cleaners, deodorants, "beauty" products, and the like. This Appendix supplies you with a list of such potentially dangerous and toxic substances.

If you can find no substitute and must use the substances listed below, at least wear rubber gloves and/or a breathing mask. Also, be sure that the area is adequately ventilated.

I urge you to "detoxify" your home. Go through each room, beginning with the kitchen, and remove all toxicants, all substances on the following list. You may be surprised how enlightening reading a few labels can be, how dangerous your life has been, and how healthy it can be from now on.

COMMERCIAL ALL-PURPOSE CLEANERS

Avoid complex phosphates, chlorinated phosphates, morpholine, petroleum-based surfactants, dry bleach, kerosene, sodium bromide, glycol ether, Stoddard solvent, EDTA, and naphtha.

Chlorinated materials form organ-chlorine compounds and are stored in fat cells that can enter mothers' milk. Morpholine and glycol are potential liver and kidney toxins. Glycol ether, Stoddard solvent, naphtha, and kerosene are neurotoxins that can cause confusion, headaches, lack of concentration, and other mental symptoms.

COMMERCIAL DEODORIZERS

Avoid methoxychlor, aromatic hydrocarbons, salicylates, petroleum distillates, formaldehyde, p-Dichlorobenzene, piperonal butxide, o-phenylphenl, and naphthalene.

Methoxychlor, dichlorobenzene, aromatic hydrocarbons, and naphthalene are potential neurotoxins, while salicylates may cause strong allergic or toxic reactions. Formaldehyde and piperonal butoxide are potential carcinogens.

COMMERCIAL DISH DETERGENTS

Avoid petroleum-based surfactants, naphtha, chloro-o-phenylphenol, germicides, diethanolamine, complex phosphates, and sodium nitrates. Chloro-phenylphenol is toxic, while diethanolamine is a potential liver toxin. Naphtha is a neurotoxicant.

COMMERCIAL DISINFECTANTS

Do not purchase or use substances containing the following: naphtha, butyl cellosolve, chlorinated germicides, petroleum-based surfactants, sodium hypochlorite, sodium sulfite, or nitrite.

Naphtha is a neurotoxicant while butyl cellosolve and sodium nitrite is strong toxins.

COMMERCIAL FURNITURE POLISHES

Avoid petroleum-distillates, propellants, diglycol laurate, amyl acetate, petroleum-based waxes, and mineral spirits. Diglycol laurate, amyl acetate, and mineral spirits are neurotoxins, while diglycol laurate is a potential liver and kidney poison. Mineral spirits contain the carcinogen benzene, and may cause lung and sinus irritation.

COMMERCIAL GLASS CLEANERS

Avoid organic solvents, petroleum-based waxes, complex phosphates, ammonia, phosphoric acid, alkyl phenoxy polyethoxy ethanols, naphtha, and butyl cellosolve.

Organic solvents, naphtha, and petroleum-based waxes are neurotoxins. When using organic solvents, beware of carcinogens such as benzene. Butyl cellosolve is a potential toxin. Phosphoric acid and ammonia are irritating and may disrupt DNA stability.

COMMERCIAL LAUNDRY DETERGENTS

Avoid petroleum-based surfactants of the aryl and alkyl group, tetra potassium pyrophospate, complex phosphates, fluosilicate, sodium toluene, xylene sulfonate, EDTA, optical brighteners, and benzethonium chloride. Tetra potassium pyrophosphate is irritating and toxic while fluosilicate is a toxic pesticide. Benzethonium chloride is potentially toxic.

COMMERCIAL METAL POLISHERS

Avoid perchloroethylene, chromic acid, plasticizers, silver nitrate, phenolic derivative, kerosene, synthetic waxes, chromic acid, naphtha, and other organic solvents.

Perchloroethylene, kerosene, naphtha, chromic acid, and organic solvents are neurotoxins. Perchloroethylene is a potential carcinogen and is toxic to the kidneys and liver; exposure may be fatal. Silver nitrate is highly toxic and corrosive. Chromic acid is a liver and kidney toxin, and a possible carcinogen.

COMMERCIAL OVEN CLEANERS

Avoid ether-type solvents, petroleum distillates, methylene chloride, butyl cellosolve, and lye.

All of the above ingredients except lye are neurotoxins. Methylene chloride is a chlorinated hydrocarbon, which is stored in fatty tissue, and is a liver and kidney toxin. Lye is a corrosive poison. Ether-type solvents commonly contain the carcinogen benzene, which may lead to respiratory symptoms.

COMMERCIAL SPOT REMOVERS

Avoid p-hydroxybenzoic acid, oxalic acid, naphtha, benzene, perchloroethylene or trichloroethylene, sodium hypochlorite, hydrofluoric acid, aromatic petroleum solvents, aliphatic hydrocarbons, chlorinated hydrocarbons, or other petroleum hydrocarbons.

The ingredients listed above are extremely toxic to many parts of the body and contain suspected and known carcinogens, exposure to which can be fatal.

COMMERCIAL TOILET BOWL CLEANERS

Avoid complex phosphates, o- or p-Dichlorobenzene, chlorinated phenols, kerosene, salicylates, germicides, fungicides, 1, 3-Diochloro-5, sodium acid oxalate, and sodium acid sulfate.

Sodium acid oxalate, chlorinated phenols, and o- or p-Dichlorobenzene are highly toxic. Sodium acid sulfate is highly irritating and corrosive. Chlorinated phenols are corrosive, metabolic stimulants. Fungicides and germicides can be toxic and cause liver and kidney damage. O- or p-Dichlorobenzene is a liver and kidney poison and neurotoxicant.

REMEMBER...

Do not purchase these toxicants in the first place. Go through your home and detoxify each room, one room at a time. Wrap each dangerous substance in newspaper and discard it carefully. Believe me, you and your family will be healthier and happier without them. However, if you must use these substances, take precautions such as wearing gloves, donning a breathing mask, and opening windows for adequate ventilation. Never take chances with your health.

Manufacturers and Suppliers of Commercial Household Products with Low Toxicity

Arm & Hammer -Arm & Hammer Baking Soda
Church & Dwight CO. Inc. Dept CG
P.O. Box 7648, Princeton, NJ 08543-7648Products available in supermarkets.

Bon Ami Company Faultless Starch/Bon Ami Co., Dept CG
1025 West 8th St. Kansas City, MO 64101 Products available in supermarkets.

Earthrite Dept CG Corp. Center # 1 55 Federal Rd. Danbury, CT 06813 203-731-5000
Nontoxic laundry detergents. Products available in health food stores.

E Magazine 28 Knight Street, Norwalk, CT 06881
An extraordinary magazine focusing on toxic environmental issues. Offers safe, nontoxic product ads, information.

EcoMall P.O. Box 20553, Cherokee Station, NY 10021 212-535-1876

Features hundreds of environmentally friendly companies and products.

Ecoshop, Inc 5884 E. 82nd St.Indianapolis, IN 46250 317-84-WORLD

Nontoxic household cleaners for home and office. Products available in health food stores.

Infinity Herbal Products Dept. CG Toronto, Canada M3J 3J9

Maker of Heavenly Horsetail All-Purpose Cleaner. Products available in health food stores.

Seventh Generation, Dept CG Colchester, VT 05446-1672 800-456-1198

Low toxicity products for household cleaning. Products available in health food stores.

Recommended Reading for Keeping Your Home Nontoxic

Check the Internet for sources where you may purchase these valuable books, or consult your local library to find a copy. It will be well worth your time and effort.

- ☐ Dadd, Debra Lynn, and Jeremy P. Tarcher. The Nontoxic Home: Protecting Yourself and Your Family from Everyday Toxins and Health Hazards. Revised edition, Ceres Press, Woodstock, NY, 1992.

- ☐ Dadd, Debra Lynn, and Jeremy P. Tarcher. Nontoxic and Natural: A Guide for Consumers; How to Avoid Dangerous Everyday Products and Buy or Make Safe Ones. Ceres Press, Woodstock, NY, 1990.

- ☐ Garland, Ann Witte. For Our Kid's Sake: How to Protect Your Child Against Pesticides in Food, New York, Mothers and Others for Pesticide Limits and the National Resources Defense Council, 1989.
- ☐ Gosselin Robert, M.D., Ph.D., Roger P. Smith, Ph.D., Harold C. Holdge, Ph.D., and Jeanne Braddock. Clinical Toxicology of Commercial Products, 5th edition. Baltimore, MD: Williams and Wikins, 1984.
- ☐ Bond-Berthold, Anne. Clean & Green. Woodstock, NY: Ceres Press, 1994.

Appendix B – Posture & Exercise: The Key to Total Body Alignment

Obese people stand crooked even when they believe they are standing straight as an arrow. If you think your posture is great, try the following simple balance test:

- Stand on one foot for 30 seconds (do both sides). If you cannot balance on one foot for 30 seconds, or if you veer from side to side, not staying perfectly balanced, your posture is not where it should be for optimal nervous system functions. Your body is not aligned. This means that when you do walk or exercise, your motion is not symmetrical, and your body is compensating by working some muscles more and others less. When one leg is worse than the other, there is unbalanced muscle function, which causes subluxations, spinal fixations, and joint stress.

Despite the availability of hundreds of structural alignment therapies and many different chiropractic and osteopathic adjustment techniques, the overwhelming majority of patients have weak balance on one or both sides. A stressed posture requires more energy for the body to stay balanced and does not allow organ-spinal connections to fully heal and regenerate.

Weak and protruded abdominal muscles cause stubborn spinal fixations, with resultant lumbar (lower back) tension and neck (cervical) weaknesses. Since millions of nerve impulses per second depend on good posture to work efficiently, it is not surprising that these neural inhibitions can make it hard to get healthy or lose weight and to properly realign the body.

While leg raises, sit-ups, and weight training exercises have abdominal strengthening effects, they commonly worsen spinal fixations and result in poor posture, rounded shoulders, rock-hard hip flexors and neck extensors, and a concave chest. An excellent way to counteract these side effects is to link walking with postural changes that engage and strengthen the deeper *transverse abdominis* muscles that stabilize the lower back (lumbar spine) and the internal and external oblique muscles that allow bending and twisting to the side.

The abdominal muscles keep the organs in their proper position. The *rectus abdominis* goes up and down the body, while the transverse is underneath going crosswise. If the rectus or transverse abdominals are out of balance with one another, they pull the spine out of alignment and weaken the digestive organs.

This means one must stop doing crunch-type sit-ups and work on overall posture to correct spinal fixations that inhibit digestion and nervous system function. Posture is how you balance your body, and only those with a perfect body have good posture. The goal is to achieve the best alignment of the entire body. Posture is often a trade-off between flexibility and stability and between motion and effort. To accomplish good posture, you have to control your body's position, keeping your body upright and stable while walking.

Since this requires more energy and continuous effort, many find these techniques formidable at first. To minimize these initial reactions, start this technique for only 5 minutes a day. Then, increase the time 5 minutes daily until you are doing 30 minutes a day. Thereafter, retraining weak muscles is best accomplished with 30 minutes a day of fast walking that includes the following postural efforts:

1. Walk tall with your head high and toward the sky.
2. Keep your shoulder blades back and chest forward.
3. Pull your belly button inward (toward your spine). Do not tilt your pelvic forward or backwards. And, always walk with good arch support.

People tell me that practicing these three simple steps while walking allows them to breathe deeper and have less stress and more energy. Stronger posture provides fast relief from stubborn back or neck tension or pain. Patients also report improvement with problems ranging from knee pain to headaches, and report feeling and looking taller. Instruct your patient to do this exercise daily and to walk like a winner, taking deep breathes at periodic intervals. These postural exercises help to stretch shortened and tense muscles and ligaments, giving the body more flexibility. In addition, doing stretching after the exercise also helps to increase flexibility and range of motion for the joints. This helps to keep the body in better balance and less prone to injury from falls. Stretching can be done anytime, anywhere — in your home, at work, or when you're traveling.

THE BOTTOM LINE - Posture is the position in which the body is held upright against gravity while walking, standing, sitting, or lying down. Exercise, combined with retraining the body's posture and eating a healthy diet with Quorum Nutrition™, can keep the body energized and in a rapid fat-burning mode. As an added bonus, retraining your body's posture helps you achieve total body alignment and greater range of motion and flexibility. In this manner strain on your supporting muscles and ligaments is minimized. Additionally, you help prevent strain, backaches, and muscular pain, as well as decrease the tension and stress on the ligaments that keep your spine properly aligned. Finding a healthy balance of diet and postural exercise is a vital step in winning the battle against obesity and degenerative disease.

Appendix-C: Advanced Commensal Cells Research to Share with your Doctor

Stabilization and Novel Targeted Delivery of Probiotic Commensals--Dr. Paul Yanick

Probiotics are live commensal cells that have beneficial effects in the amelioration or prevention of both acute and chronic health conditions. Commensals—more than a trillion per every gram of intestinal tissue—are immunological peacekeepers that promote the digestion of food and destroy dangerous microbes. Two Nobel laureates have exploited commensal flora as a realistic therapeutic strategy for infectious, inflammatory and neoplastic disorders.18-20 Commensals are essential health assets that confer protection against infections, prime mucosal immunity, produce a rich repository of nutrients and metabolites, and that maintain the reciprocity of immune and genetic mechanisms (SIRT1). They have confirmed efficacy in acute enteric infections, post-antibiotic syndromes, colitis, irritable bowel syndrome, and are critical for achieving the goal of immunorestoration,17-29 and can prevent or treating acute conditions2, 5-14, manage irritable bowel syndrome,12,16, 69,70 lactose intolereance,7,4 chronic liver disease,11 pancreatitis,9 and certain forms of cancer.6, 16,17,56

Only an estimated 5-10% of the cells within your body are your own[5], the remaining 90-95%

are commensal cells. These microorganisms thrive when nourished with pre- and synbiotic nutrients and perform a number of vital functions. Since the balance of the constituent strains can have a profound effect on the intestinal well-being of their host, as anyone who suffered from microbial food poisoning can testify, the clinical goal is to RESTORE the balance of commensal organisms and make sure that this microbial teamwork can dominant potential pathogens. As you will learn, even too much of certain good probiotics can have extremely negative effects on the overall function of the body's commensals. Hence for the last deacde my research has targeting the microbe-microbe ratios that yield the best overall probiotic effect.

In order to be effective probiotics have to mimic the competitive interactions, antagonism of pathogens, and production of synbiotic nutrients and anti-microbial factors of commensal cells which are known as the most renewable metabolic organ of the body with a level of metabolic activity comparable to the liver. The metabolic repertoire of the flora includes the synthesis of active synbiotic compounds that create a gut immunological barrier. A *synbiotic* is defined as a nanoscale nutrient that is formed from a mixture of probiotics and prebiotics and can RESTORE commensal cell activity in a wide spectrum of clinical disorders.[30-39]

However, there is vast need for improving shelf-life and packaging probiotics in a format that will ensure they find a home in the human gut ecosystem. My research has focussed on the maintenance of the activity of probiotic-commensal cells during processing and storage as well as their targeted delivery to the lower intestine. Progress with the clinical use of probiotic and synbiotics has been delayed because manufacturers often fail to:

- **Eliminate mycotoxins** - our research and that of other scientists has documented high levels of mycotoxins (mold) in a high percentage of cultured or fermented probiotic or synbiotic products.[40-43] For example, in one university study of 49 samples, 41 had dangerous levels of fungal isolates.[40]

- **Maintain Quantity Control -** a bewildering array of soft claims and standardized verification of product bioactivity, composition, stability and shelf life result in formulations with microbe-microbe competition and that do not mimic host-flora signaling pathways.

- **Exploit Host-Commensal Signaling Pathways** – According to a leading commensal researcher at the University of Ireland "*mucosal homeostasis requires continual signaling from bacteria within the lumen of the gut. It is a question of mimicking the flora and exploiting host-flora signaling pathways*." Researchers now understand that host-flora signaling is a function of *riboswitches* and pattern recognition receptors that attenuate inflammatory responses and allow commensals to take residence in the gut. Therefore, a mineral-ligand matrix--critical to *riboswitch signaling* of commensals—should logically be included in probiotic formulations. Since Yale University shows that commensals make synbiotic nutrients with *riboswitches,*[44, 49-51] finding ways to exploit these signaling pathways in clinical practice may yield superior results with probiotics.

- **Mimic Gut Flora** - Competitive microbe-microbe interactions in probiotics must mimic commensal flora in the human gut. By accomplishing this feat, probiotic become commensals and are not just transient in nature. They stay in the gut long enough to maintain the physiological normal state of inflammation and activate gut-associated lymphoid tissue (GALT). GALT is the largest immune organ in the human body, resulting in the stimulation of T and B cells and the establishment of the cytokine networks. It is critical to understand that once a person has taken an antibiotic or a natural anti-infective, they only have ten percent (instead of ninety percent) population of commensals. For over a decade, I have studied how different probiotics can disrupt the reciprocity of the TH-1 and TH-2 cytokine responses and looked carefully at how some can compete against others to diminish their total effectiveness. Borrowing further from nature's design, I fermented foods to produce mineral-ligands that could turn "on" the toggle-like riboswitches for a greater range of bioactivity or "off" in pathogenic bacteria.[44-48]

- **Nourish Commensals -** Commensal cells make proteins--the wheels, cogs, chutes and conveyor belts that transport the synbiotic nutrients to cells—and, co-enzymes in small, "nano" sized molecules that cooperate with proteins and enzymes to ignite powerful biochemistry in the body. Epic metabolic pathways are activated in the construction of these nutrients which can occur in the fermentation tank or in the body as commensals construct nutrients from pre- and synbiotics and strictly control nutrient levels by shutting synthesis down when nutrients are ample and the cells infrastructure is optimal. [30-39]

While pharmaceuticals and natural anti-infective herbs can save lives, they are indiscriminate killers of all commensal cells. Antacids and digestive enzymes destroy commensals or alter commensal cell metabolism, creating a favorable environment for opportunistic yeast and fungal infections. In addition, excessive alcohol and sugar, NSAID's, radiation, chlorine or fluorine, and inorganic, non-covalent ionic minerals are extremely harmful to commensal cells.

Table 1 (below) illustrates the specific strains used in our clinical studies designed to exploit signaling pathways, improve acid- and bile-tolerance, and eliminate unwanted microbial competition. Powerful antagonistic actions against a wide spectrum of microbial pathogens can be achieved with these design criteria.

Lactobacillus acidophilus	Immune enhancer, important commensal, prevents diarrhea or constipation, anti-inflammation effects, support GALT [18-20;58-62;67]
Lactobacillus Casei	Immune enhancer, positive anti-diarrhea effects, positive immune effects with cancer [52, 55-57]
Lactobacillus plantarum	Inhibits bacterial translocation and secondary septic responses[52]
Lactobacillus Salivarius	Minimizes abdominal fat storage, anti-clostridium effects, attenuates GI inflammation, and prevents tumor formation[63-66]
Lactobacillus Rhamnosus	Prevents eczema and has anti-allergenic effects. [54, 58-62]
Bifidobaterium bifidum	Prevention of acute infectious diarrhea, anti-clostridium effects[23-29,68]
Saccharomyces boulardi	Anti-diarrhea effects, critical for mucosal immunity[71-74]

Again, immune reciprocity is contingent on commensal cell nourishment. Commensals are the immune system's primary weaponry against unwanted microbial invaders. When depleted or wiped out from antibiotics or anti-infectives (herbal, silver, hydrogen peroxide or stabilized oxygen) or synthetic, man-made chemicals they are unable to proliferate and take residence in the gut. The result: the immune system is driven into exhaustion from long-term immunological warfare and battles that cannot be won. This exhaustion, called immunosuppression, explains why runaway viral can't be controlled and why so many patients struggle with cyclic mycotic and bacterial infections.

In summary, there are a number of problems in determining the efficacy of probiotics as a whole. In order to resolve some of these issues for a specific product, our work has covered the formulation, stabilization and targeted delivery of probiotic microorganisms shown in **Table 1**. In addition, entrapment in a mineral-ligand matrix helps formulation and delivery problems of current probiotics and has a number of distinct advantages in boosting commensal cell populations in sick patients.

References

1. Chan, WA et al: *Minerva Biotechnol.*, 12:271-278.
2. Dunne, C et al: 1999. Antonie van Leeuwenhoek, 76:297-292.
3. Dunne, C et al: 2001. *Am J Clin Nutr.*, 73:386-392.
4. Friedman, G. 2000. *Gastroenterology*, 118:5179.
5. Hatakka K et al: *Br. Med. J.*, 322:1327-1329.
6. Hosono, A et al: 1986. *J. Dairy Sci.*, 2237-2242.
7. Kolars, JC et al: 1983. *Clin. Res.*, 31:764.
8. Lee, YK et al: 2000. *Appl. Environ. Microbiol.*, 66:3692-3697.
9. Mangiante, G et al: 2001. *Dig. Surg.*, 18:47-50.
10. Neeser, JR et al: 2000. *Glycobiology*, 10:1193-1199.
11. Muting, D. 1968. *Am. J. Proctol.*, 19:5.
12. Nobaek, S et al: 2000. *J. Gastroenterol.*, 95:1231-1238.
13. Pochapin, M. 2000. *Am. J. Gastroenterol.*, 95:11-13.
14. Sartor, R.B. 2001. *Curr. Op. Gastroenterology*, 17: 324-330.
15. Shortt, C. 1998. *Chem. Ind.*, April:300-303.
16. Takagi, A et al: 2001. *Carcinogenesis*, 22: 599-605.
17. Zabala, A et al; 2001. *Lett. Appl. Microbiol.*, 32:287-292.
18. Shanahan F 2005. *Am J Physiol Gastrointest Liver Physiol* 288:.
19. Shanahan F 2004. *Inflamm Bowel Dis* (Suppl 1);S16-S24.
20. Shanahan F 2004 *Adv Drug Delivery Rev* ;56:809-818.
21. Elliott DE et al: 2005 *Curr Opin Gastroenterol* 21:51-58.
22. Summers RW et al: 2005 *Gut*;54:87-90.
23. Gill HS et al: 2004 *Postgrad Med J* 80:516-526.
24. Senok AC et al: 2005. *Clin Microbiol Infect* 11:958-966.
25. Floch MH et al 2005. *Gastroenterol Clin North Am* 34:547-570.
26. Allen SJ et al: 2004 *Cochrane Database Sys Rev.* (2):CD003048.
27. D'Souza AL et al: 2002. *BMJ* ;324:1361.
28. McCarthy J et al: 2003 *Gut* 52:975-980.
29. Di Giacinto C et al: 2005. *J Immunol* 174:3237-3246.
30. Furrie E et al: 2005 *Gut* 242-249.
31. Collins MD et al: 1999 *Am J Clin Nutr* 69:1046S-51S.
32. Olah A et al: 2002 *Br J Surg* 89:1103-7.
33. Rayes, N et al 2002 *Nutrition* 18:609-615.
34. Rayes, N et al 2002. *Transplantation* 74:123-7
35. Bengmark, S. 2002: *Considerations in the Intensive Care Unit* - Kendall/Hunt Publishing Company, Iowa Chapter 34:381-399
36. Bengmark S 2002: *Current Opinion in Critical Care* 8:145-151.
37. Bengmark S. 2002: *Nutritional Considerations in the Intensive Care Unit* Kendall/Hunt Publishing Company, Iowa Chapter 33:365-380.
38. Bengmark, S. 2001 *Cur Opin Clinl Nut Metab Care* 4(6).
39. Bengmark, S. 1998 *Gut* 42; 2-7.
40. Jonsyn FE: 2005. *Mycopathologica*, 10:2.

41. Ogunsanwo BM: *Nahrung.* 1989;33(10):983-8.
42. Ogunsanwo BM *Nahrung.* 1989;33(5):485-7.
43. Ogunsanwo BM *Food Chem Toxicol.* 2004 Aug;42(8):1309-14.
44. Pornsuwan S et al: 2006. *J Am Chem Soc*, 128:12; 3892-4011.
45. Barrick JE et al: 2007. *Scientific American* 50-57.
46. Blout KF et al: 2007. *Nature Biotechnology*.
47. Sudarsan N et al: 2003. *RNA*, 9:6; 664-47.
48. Winkler W et al: 2003, *Nature* 419:952-6.
49. Collins, TJ. *Science* 291:5501: 48-49, Jan 5, 2001.
50. Marchese, SR et al. 1999 *Trends in Pharmacologcial Sciences*, 20:9, 370-75.
51. Brady AE et al. 2002 *Cellular Signaling*. 14:4; 297-309.
52. Cunningham RS et al: 2000. *Am J Gastroenterol* 95:22-25S.
53. Isolauri E et al: 2000. *Clin Exp Allergy*, 30: 1604-10.
54. Pessi T et al: 2000. *Clin Exp Allergy* 30(12):1804-8.
55. Dugas B et al. 1999. *Immunology Today*. 20(9): 387-90.
56. Rafter J. 2003 *Best Pract Res Clin Gastroenterol*.(5)849-59.
57. O'Mahony L et al: 2001. *Alim Pharm Ther*: 25:1219-25.
58. Cross ML. 2002. *FEMS Immunol Med Microbiol*; 34(4):245-53.
59. Guarner F et al: 2003. *Lancet*, 361(9356):512-9.
60. Helgeland L et al: 1996 *Immunol* 89:494-501.
61. Cebra JJ. 1999. *Am J Clin Nutr* 69:1046S-1051S.
62. Cebra JJ et al: 1998. *Dev Immunol*, 6:13-18.
63. Backhed F et al: *Proc Natl Acad Sci U S A* 2004;101:15718-23.
64. Ley RE et al: *Proc Natl Acad Sci U S A* 2005;102:11070-75.
65. Shanahan F. *Am J Physiol Gastrointest Liver Physiol* 2000;**278**.
66. Shanahan F. *Inflamm Bowel Dis* 2000;**6**:107-115
67. Lin HC et al: 2005. *Pediatrics,*115:1-4.
68. Bin-Nun A et al: 2005. *J Pediatr,*;147:192-196.
69. Verdu EF et al: 2005. *Cur Opin Gastroenterol,* 21:697-701.
70. O'Mahony L et al: 2005. *Gastroenterology,* 128:541-551.
71. McFarland, LC: 2006. *Am J Gastroenterol* 101:4, 812–822.
72. Reid, G et al: 2006 *FEMS Immunology & Medical Microbiology* 46:2, 149–157.
73. Polk, B et al: 2006. *Curr Opin Clin Nutr Metabolic Care* 9:6, 717-21.
74. Szajewska, H et al: 2006. *J Ped Gastroenterol Nutr* 42:5, 454.

My Selection of some of the most Impressive research on Commensals

Bengmark S. (2002): Aggressive peri- and intraoperative enteral nutrition - Strategy for the future. In: Shikora SA, Martindale RG, Schwaitzberg SD (eds): Nutritional Considerations in the Intensive Care Unit - Science, Rationale and Practice. Kendall/ Hunt Publishing Company, Dubuque, Iowa USA. Chapter 33:365-380

Bengmark S. (2002): Use of Pro-, Pre- and Synbiotics in the ICU - Future options. In Shikora SA, Martindale RG, Schwaitzberg SD (eds): Nutritional Considerations in the Intensive Care Unit - Science, Rationale and Practice. Kendall/Hunt Publishing Company, Dubuque, Iowa USA. Chapter 34:381-399

Frame LT, Hart RW, Leakey JEA et al. (1998): Caloric Restriction as a Mechanism Mediating Resistance to Environmental Disease. Environmental Health Perspectives 106:313-324

Bengmark S. (2001): Nutritional modulation of acute and "chronic" phase response. *Nutrition* 17:489-495

Alverdy JC, Laughlin RS, Wu R (2003): Influence of the critically ill state on host pathogen interactions within the intestine: Gut derived sepsis redefined. *Critical Care Medicine* 31:

Schook LB, et al: (1976): Murine gastrointestinal tract as a portal of entry in experimental Pseudomonas aeruginosa infections. *Infection and Immunity* 14:564-570

Kurahashi K, Kajikawa O, Sawa T et al. (1999): Pathogenesis of septic shock in Pseudomonas aeruginosa pneumonia. *Journal of Clinical Investigation* 104:743-750

Kinney KS, Austin CE, Morton DS et al. (2000): Norepinephrine as a growth stimulating factor in bacteria: Mechanistic studies. *Life Science* 67:3075-3085

McWhirter JP, Pennington CR (1994): Incidence and recognition of malnutrition in hospital. *British Journal Medicine* 308:945-948.

Giner M, et al:(1996): In 1995 a correlation between malnutrition and poor outcome in critically ill patients still exists. *Nutrition* 12:23-29

Pikul J, Sharpe MD, Lowndes R, Ghent CN (1994): Degree of preoperative malnutrition is predictive of postoperative morbidity and mortality in liver transplant recipients. *Transplantation* 57:469-472

Stoutenbeck CP, et al: (1984): The effect of selective decontamination of the digestive tract on colonization and infection rate in multiple trauma patients. *Intensive Care Medicine* 10:185-192

Nathens AB, Marshall JC (1999): Selective decontamination of the digestive tract in surgical patients: a systemic review of the evidence. *Archives of Surgery* 134:170-176

D'Amico R, Pifferi S, Leonetti C et al. (1998): Effectiveness of antibiotic prophylaxis in critically ill adult patients: systemic review of randomized controlled trials. *British Medical Journal* 316:1275-1285

Hellinger WC, Yao JD, Alvarez S et al. (2002): A randomized, prospective, double-blinded evaluation of selective bowel decontamination in liver transplantation. *Transplantation* 73:1904-1909

Zwaveling JH, Maring JK, Klompmaker J et al. (2002): Selective decontamination of the digestive tract to prevent postoperative infections: A randomized placebo-controlled trial in liver transplant patients. *Critical Care Medicine* 30:1204-1209

Bengmark S, et al: (2001): Uninterrupted perioperative enteral nutrition. *Clinical Nutrition* 20:11-19

Andresen AFR (1918): Immediate jejunal feeding after gastroenterostomy. *Annals of Surgery* 67:565-566

Marik PE, Zaloga GP (2001): Early enteral nutrition in acutely ill patients. Critical Care Medicine 29:2264-2270 .

Kompan L et al: (1999): Effects of early enteral nutrition on intestinal permeability and the development of multiple organ failure after multiple injury. *Intensive Care Medicine* 25:157-161

Brandzaeg P, Halstensen TS, Krajci P et al. (1989): Immunobiology and immunopathology of human gut mucosa: Humoral immunity and intraepithelial lymphocytes. *Gastroenterlogy* 97:1562-1584

Kiyono H, McGhee JR (1994): T helper cells for mucosal immune responses. In Ogra, P.,L., Mestecky, J., Lamm,

M.E., et al. (eds): Handbook of mucosal immunology, Orlando, Florida, Academic Press 263-274

Shirabe K, et al: (1997): A comparison of parenteral hyperalimentation and early enteral feeding regarding systemic immunity after major hepatic resection. *Hepato-Gastroenterology* 44:205-209

Windsor ACJ et al:. (1998): Compared with parenteral nutrition, enteral feeding attenuates the acute phase response, and improves disease severity in acute pancreatitis. *Gut* 42:431-435

Gnoth MJ, et al: (2000): Human milk oligosacharides are minimally digested in vitro. *Journal of Nutrition* 130:3014-3020.

De Felippe JJ, et al: (1993): Infection prevention in patients with severe multiple trauma with the immunomodulator 1-3 polyglucose (glucan). *Surgery, Gynecology, Obstetrics* 177:383-388

Rabbani GH, Teka T, Zaman B et al. (2001): Clinical studies in persistant diarrhea; dietary management with green banana or pectin in Bangladesh children. Gastroenterology 121:554-560

Olasupo NA, Olukoya DK, Odunfa SA (1995): Studies on bacteriocinogenic Lactobacillus isolates from selected Nigerian fermented foods. *Journal of Basic Microbiology* 35:319-324

Finegold SM, Sutter VL, Mathisen GE (1983): Normal indigenous intestinal flora. In Hentges DJ (ed). Human intestinal microflora in health and disease, London, Academic Press 3-31

Ahrné S, Nobaek S, Jeppsson BG et al. (1998): The normal Lactobacillus flora of healthy human rectal and oral mucosa. *Journal of Applied Microbiology* 85:88-94

Bennet R, Nord CE (1987): Development of the faecal anaerobic microflora after caesarean section and treatment with antibiotics in newborn infants. *Infection* 15:332-336

Johansson ML, Molin G, Jeppsson B et al. (1993): Administration of different lactobacillus strains in fermented oatmeal soup: in vivo colonization of human intestinal mucosa and effect on the indigenous flora. *Applied Environmental Microbiology* 59:15-20

Bengmark S (2000): Prospect for a new and rediscovered form of therapy: Probiotic and phage. In Andrew PW et al (eds) Fighting Infection in the 21st century. Blackwells, London 97-132

Isenmann R, Rau B, Beger HG (1999): Bacterial infection and extent of necrosis are determinants of organ failure in patients with acute necrotizing pancreatitis. *Brit J Surg* 86:1020-1024

Beger HG, Bittner R, Büchler MW (1986): Bacterial contamination of pancreatic necrosis - a prospective clinical study. *Gastroenterology* 91:433-438

Büchler MW, Gloor B, Müller CA et al. (2000): Acute necrotizing pancreatitis: treatment strategy according to the status of infection. *Annals of Surgery* 232:619-626

Kingsnorth A (1997): Role of cytokines and their inhibitors in acute pancreatitis. *Gut* 40:1-4

Qamruddin AO, Chadwick PR (2000): Preventing pancreatic infection in acute pancreatitis. *Journal of Hospital Infections* 44:243-253

Golub R, Siddiqi F, Pohl D (1998): Role of antibiotics in acute pancreatitis: a meta-analysis. Journal of *Gastrointestinal Surgery* 2:496-503

Oláh A, Belágyi T, Issekutz Á et al. (2002): Early Enteral Nutrition with Specific Lactobacillus and Fibre reduces Sepsis in Severe Acute Pancreatitis. *British Journal of Surgery* 89:1103-1107

Rayes N, Hansen S, Boucsein K et al. (2002): Early enteral supply of fibre and lactobacilli vs parenteral nutrition - a controlled trial in major abdominal surgery patients. *Nutrition* 18:609-615

McNaught CE et al: (2002): A prospective randomized study of the probiotic Lactobacillus on plantarum 299V on indices of gut barrier function in elective surgical patients. *Gut* 51:827-831

RayesNetal:(2002):EarlyenteralsupplyofLactobacillus and fibre vs selective bowel decontamination (SBD) - a controlled trial in liver transplant recipients. *Transplantation* 74:123-127

Kruszewska D, Lan JG, Lorca G et al.: (2009) Selection of lactic acid bacteria as probiotic strains by in vitro tests. *Microbial Ecology in Health and Disease*. .

Ljungh °A, Lan J-G, Yamagisawa N (2002): Isolation, selection and characteristics of Lactobacuillus paracasei ssp paracasei isolate F19. *Microbial Ecology in Health and Disease* 3:4-6

Hoyos AB (1999): Reduced incidence of necrotizing enterocolitis associated with enteral administration of Lactobacillus acidophilus and Bifidobacterium infantis to neonates in an intensive care unit. *International Journal of Infectious Diseases* 3:197-202

Buscher HJ et al: (1999): Preliminary observations on influence of diary products on biofilm removal from silicon rubber voice prostheses in vitro. *Journal of Dairy Science* 83:641-647

Van der Mei HC et al: (2000): Effect of probiotic bacteria on prevalence of yeasts in oropharyngeal biofilms on silicone rubber voice prostheses in vitro. *Journal of Medical Microbiology* 49:713-718

Bengmark S (2002): Gut microbial ecology in critical illness: is there a role for pre-, pro-, and synbiotics. *Current Opinion in Critical Care* 8:145-151.

Curr Opin Biotechnol, 2001 Oct, 12(5), 499 - 502 -**Food processing: probiotic microorganisms for beneficial foods**; Schiffrin EJ et al.; Human studies have demonstrated that selected probiotic strains can influence the composition of the intestinal **microflora** and modulate the host immune system with promise demonstrated for the application of probiotics in human disease.

Drug Metab Dispos, 2001 Nov, 29(11), 1440 - 5 -**Identification of new derivatives of sinigrin and glucotropaeolin produced by the human digestive microflora using 1H NMR spectroscopy analysis of in vitro incubations**; Combourieu B et al.; One- and two-dimensional (1)H NMR spectroscopy were used to study commesnal cell microflora.

Rev Argent Microbiol, 2001 Jul-Sep, 33(3), 133 – 40 -**{Clinical and microbiological study of adult periodontal disease}**; Nogueira Moreira A et al - Significant differences in the subgingival **microflora** between healthy and disease sites in patients with moderate and severe periodontitis.

Carcinogenesis, 2001 Oct, 22(10), 1721 – 5 - **Intestinal microflora plays a crucial role in the genotoxicity of the cooked food mutagen 2-amino-3-methylimidazo {4,5-f}quinoline**; Kassie F et al.; The role of intestinal bacteria on health risks caused by dietary carcinogens.

Acta Odontol Scand, 2001 Aug, 59(4), 244 – 7 -**Regulation of experimental mucosal inflammation**; Strober W et al.; Studies conducted over the past 10 years have provided ample evidence that many types of inflammations arising from basic abnormalities of immune regulation are ultimately 'funneled' through a Th1 or Th2 T cell-mediated immune reaction . Intestinal disease related to an excessive immune response to elements of the bacterial **microflora** of the gut.

Zh Mikrobiol Epidemiol Immunobiol, 2001 Jul-Aug, (4), 95 – 7 -**Effect of radio wave-induced hyperthermia on microflora of the prostate in the treatment of prostatitis associated with infertility**; Kuz'min MD et al.; Microflora and prostate disease.

Zh Mikrobiol Epidemiol Immunobiol, 2001 Jul-Aug, (4), 84 – 6 - **{Spectrum of microflora isolated from various areas of the female reproductive tract}**;

Deriabin DG et al.; The composition of **microflora** in female reproductive tract.

Zh Mikrobiol Epidemiol Immunobiol, 2001 May-Jun, (3), 76 – 80 - **Effect of the lavage of the digestive tract on microflora in patients with polyps in the large intestine**; Korshunov VM et al.; The microbial status of the intestine and the influence of dysbiotic conditions. *Dig Dis, 2001, 19(2), 144 – 7* - **Helicobacter pylori treatment: a role for probiotics?** Commensal probiotics to improve Helicobacter pylori eradication rate.

Zh Mikrobiol Epidemiol Immunobiol, 2001 Mar-Apr, (2), 57 – 61- **Qualitative composition of the normal intestinal microflora in individuals from the various age groups**; Korshunov VM et al.; The study of the **microflora** of the large intestine in healthy adult volunteers of different age groups (25-36, 55-68 and 88-94 years old).

Life Support Biosph Sci, 1999, 6(3), 193 – 7 - **Experimental microcosms as models of natural ecosystems for monitoring survival of genetically modified microorganism**; Popova LYu et al.; An experimental approach for investigation of genetically modified microorganisms (GMMO) introduced into model ecosystems to evaluate potential risk of propagation of recombinant plasmids in surrounding medium has been developed.

Klin Med (Mosk), 2001, 79(6), 39 – 41- **Impact of the impaired intestinal microflora on the course of acne vulgaris**; Volkova LA et al.; The paper deals with studies of the intestinal **microflora** in 114 patients with acne vulgaris and dysbiosis.

Lik Sprava, 2001 Mar-Apr, (2), 84 – 6 - **Microbial intestinal disturbances in children with frequent acute complicated pneumonia**; Shamsiev FM; In the investigation designed to study intestinal **microflora** in children with acute complicated pneumonia related to dysbiosis.

Vopr Pitan, 2001, 70(3), 6 – 8 - **Microbiological aspect of balanced nutrition**; Kuiarov AV et al.; Analysis of modern resources of normal human **microflora** correction has shown an importance of microbiological aspect in principles of a balanced feeding for realization of the optimum mechanism of normal human **microflora** maintenance.

J J Pediatr Gastroenterol Nutr, 2001 May, 32(5), 534 – 41 -**Gas production by feces of infants**; Jiang T et al.; BACKGROUND: Intestinal gas is thought to be the cause abdominal discomfort in infants and dysbiosis.

Am J Physiol Gastrointest Liver Physiol, 2001 Jul, 281(1), G144 – 50 - **Mechanism of thiamine uptake by human colonocytes: studies with cultured colonic epithelial cell line NCM460**; Said HM et al.; Vitamins made by commensal microflora.

Inflamm Bowel Dis, 2001 May, 7(2), 136 – 45 - **Adaptation of bacteria to the intestinal niche: probiotics and gut disorder**; Dunne C; The gastrointestinal tract is a complex ecosystem host to a diverse and highly evolved microbial community composed of hundreds of different microbial species.

Am J Clin Nutr, 2001 Jun, 73(6), 1094 - 100 - **Determinants of serum enterolactone concentration**; Kilkkinen A et al.; BACKGROUND: The lignan enterolactone produced by the commensal microflora may protect against hormone-dependent cancers and cardiovascular diseases.

Microbiol Res, 2001, 156(1), 83 - 6 -**Epiphytic microflora of poplar clones susceptible and resistant to infection by** *Infect Immun, 2001 Jun, 69(6), 3719 - 27* -.

Selective enhancement of systemic Th1 immunity in immunologically immature rats with an orally administered bacterial extract; Bowman LM et al.; Commensal role in immune response.

Med Hypotheses, 2001 Apr, 56(4), 448 - 50 - **Redefining 'self': the role of microflora (commensals) mismatch in the development of GvHD after allogeneic stem cell transplantation and some possible remedies**; Singh HP et al.; *J Agric Food Chem, 2001 Apr, 49(4), 1751 - 60* - **Screening of intestinal microflora for effective probiotic bacteria**; O'Sullivan DJ; Evidience of sterile and ineffective probiotics.

Gastroenterology, 2001 Feb, 120(3), 622 – 35 - **Inflammatory bowel disease: immunodiagnostics, immunotherapeutics, and ecotherapeutics**; Shanahan F; IBD the result of defective regulation of mucosal immune interactions with commensal microflora.

Am J Clin Nutr, 2001 Feb, 73(2 Suppl), 444S - 450S -**Probiotics: effects on immunity**; Isolauri E et al.;

The gastrointestinal tract functions as a barrier against antigens from microorganisms and food . The generation of immunophysiologic regulation in the gut depends on the establishment of indigenous commensal microflora. *Mikrobiol Z, 2000 May-Jun, 62(3), 26 - 35-* **{The effect of antibiotic preparations and their combinations with probiotics on the intestinal microflora of mice}**; Furzikova TM et al.; Antibiotic drugs (biseptol, polymyxin, canamycin) as well as their combinations with probiotics biosporin and subalin have been studied for their effect on mice intestine commensal microflora.

Adv Microb Physiol, 2000, 42, 25 - 46 -**The intestinal microflora: potentially fertile ground for microbial physiologists**; Tannock GW; The intestinal **microflora** provides opportunities for microbial physiological research . The metabolic interactions of bacterial inhabitants of the intestinal community, bacterial bioenergetics, preferential utilization of substrates as energy sources by specific bacterial species, and intercellular signalling are among the topics of challenging research awaiting the attention of microbial physiologists.

Zh Mikrobiol Epidemiol Immunobiol, 1999 Jul-Aug, (4), 70 - 4 - **The immuno-microbiological characteristics of the small intestine and the translocation of the enteral microflora in acute intestinal obstruction**; Chernov VN et al.; Barrier for symbiotic microflora revieved and effects on body as a whole.

Inflamm Bowel Dis, 2000 May, 6(2), 107 - 15 - **Probiotics and inflammatory bowel disease: is**

there a scientific rationale? Shanahan F. - Most conventional forms of drug therapy suppress or modify the host immunoinflammatory response and neglect the other contributor to disease pathogenesis-the environmental **microflora** . Probiotics are live microbial food ingredients that alter the enteric **microflora** and have a beneficial effect on health. The rationale for using probiotics in IBD is mainly based on evidence from human studies and experimental animal models implicating intestinal bacteria in the pathogenesis of these disorders . The relationship between bacteria and intestinal inflammation is complex and does not appear to reflect a simple cause and effect . Similarly, the field of probiotics is complex and in need of rigorous research . Until the indigenous flora are better characterized and mechanisms of probiotic action defined, the promise of probiotics in IBD is unlikely to be fulfilled . Because of strain-specific variability and clinical and therapeutic heterogeneity within Crohn's disease and ulcerative colitis, it cannot be assumed that a given probiotic is equally suitable for all individuals . Although preliminary results of probiotic therapy in animal models and humans with ulcerative colitis and pouchitis have been encouraging, their efficacy in treatment or maintenance of remission of Crohn's disease remains to be clarified . However, the circumstantial evidence for some form of biotherapeutic modification of the enteric flora in Crohn's disease seems compelling . In the future, **probiotics may offer a simple adjunct to conventional therapy with the emphasis on diet shifting from one of nutritional replenishment alone to a more functional role.**

G Chir, 2000 Apr, 21(4), 196 - 204

Eur J Gastroenterol Hepatol, 2000 Mar, 12(3), 267 – 73 - **The role of the resident intestinal flora in acute and chronic dextran sulfate sodium-induced colitis in mice**; Hans W et al.; OBJECTIVE: There is increasing evidence that the intestinal **microflora** plays an important role in the pathogenesis of inflammatory bowel disease. In acute DSS-induced colitis **bacteria and/or bacterial products play a major role in initiation of inflammation but not in chronic DSS colitis**.

Ugeskr Laeger, 2000 Mar 6, 162(10), 1361 – 6 **Chronic inflammatory bowel disease--pathogenic concepts and therapeutic perspectives**; Madsen JR; Chronic inflammatory bowel disease (IBD) is considered to be a consequence of inappropriate upregulation of immune reactions evoked by dysbiosis.

Klin Med (Mosk), 2000, 78(2), 26 – 30 -**Dysbiotic conditions in patients operated and reoperated for heart defects and ischemic heart disease}**; Litasova EE et al.;Advanced dysbiosis--severe persistent condition with permanent source of endogenic infection complicated the underlying disease and bringing postoperative septic complications. *Curr Opin Clin Nutr Metab Care, 1999 Nov, 2(6), 481 – 4* - **Methods for assessing the potential of prebiotics and probiotics**; Rycroft CE et al.; Prebiotics and **probiotics are microflora management tools designed to improve human health**. Both are dietary materials that fortify components of the gut flora seen as 'beneficial'. Gut flora modulation is an important area of the nutritional sciences, however, it is imperative that reliable methodologies be used to determine efficacy. This review will discuss the current techniques used in prebiotic and probiotic research.

Mikrobiol Z, 1999 Sep-Oct, 61(5), 85 - 96 -**The interrelation of microbial ecosystems and human immunity**; Fedorovskaia EA et al.; An analytical survey of literature on the problem of the study of interrelations between the immune status of the organism and microbial ecology of a man is presented.

Appendix-D: References

In this section, you will find research notes or references. They are listed here to assist you who wish to investigate the concepts and ideas in greater depth. They are not listed in order of importance but generally follow the logic of the book. Moreover, they do not include over three decades of research used in the development of the concepts presented in this book as there is only so much information that fits comfortably in a book. I have provided as much information as possible, so you should be able to find these resources through the internet, at your local public library, and/or through the resources of college/university libraries in your area. I start with my own orginal research publications many that are available as e-publications at www.quantafoods.com.com.

- ☐ Yanick, P., and R. Jaffee. *Clinical Chemistry and Nutrition: A Physician's Desk Reference*. Coldbrook, VT: Biological Energetic Press, 1988.
- ☐ Yanick, P. "Physiological-Chemical Assessment of Undernutrition." *Townsend Letter for Doctors*, July, 1988.
- ☐ Yanick, P. "Dietary and Lifestyle Influences on Cochlear Disorders and Biochemical Status: A Twelve-month Study." *Journal of Applied Nutrition* Vol. 40, no. 2, (1988).
- ☐ Yanick, P. *Manual of Neurohormonal Regulation*. Coldbrook, VT: Biological Energetic Press, 1992.

- ☐ Yanick, P. "Experiments with Water during Various Detoxification Protocols." Unpublished Manuscript, 2001.
- ☐ Yanick P. Detoxification Breakthroughs for Allergies and Chronic Toxicity." *Townsend Letter for Doctors and Patients*. July 2001.
- ☐ Yanick P. "Meridian/Organ Nutraceutic Resonant Complexes: New Hope for Chronically-Sick Individuals." *Townsend Letter for Doctors & Patients* (May, 2000): 136–39.
- ☐ Yanick, P. "Boosting Nutrient Uptake in Chronic Illness." *Townsend Letter for Doctors & Patients*, December 2000.
- ☐ Yanick, P. "Functional Medicine Update." *Townsend Letter for Doctors*, Feb., 1994.
- ☐ Yanick, P. *Bioregulation, Regeneration and Lifespan Extension*. Yanick, Inc. 1994.
- ☐ Yanick, P. "New Insights into Brain Fog, Memory & Learning Disorders, Insomnia, Anxiety, Depression and Immune Disorders." *Townsend Letter for Doctors & Patients* (June, 2000): 154–56.
- ☐ Yanick, P. "Hormone Resistance & Ground Regulation System." *Townsend Letter for Doctors & Patients* (January 1999): 88–90.
- ☐ Yanick, P. *Quantum Medicine*. Portland, OR: Writer Service Publications, 2000.

- ☐ Yanick, P. *A Professional's Guidebook of Quantum Medicine*. Las Vegas, NV: American Academy of Quantum Medicine, 2001.
- ☐ Yanick, P., and V. Giampapa. ProHormone Nutrition. Montclair, NJ: Longevity Institute International, 1998
- ☐ Yanick, P., and V. Giampapa. *Quantum Longevity*. Los Angeles, CA: Promotion Publishing, 1997.
- ☐ Yanick P. "Lymphatic Therapy for Chronic Immune & Metabolic Disorders, Detoxification and Successful Pain Elimination." *Townsend Letter for Doctors* (January 1995): 34–40.
- ☐ Yanick, P. "Food Supplement Benefits and Risks in Carcinogenesis: Part I." *Townsend Letter for Doctors & Patients* (Oct 2001).
- ☐ Yanick, P. "Food Supplement Benefits and Risks in Carcinogenesis: Part II." *Townsend Letter for Doctors & Patients* (Dec 2001).
- ☐ Yanick, P. "Biomolecular Nutrition and the GI Tract." *Townsend Letter for Doctors* (Dec. 1993):1248–1250.
- ☐ Yanick, P. "Disorders of Gall Bladder & Duodenum in Overweight Patients." *Townsend Letter for Doctors* (June 1994): 568–570.
- ☐ Yanick, P. "Functional Correlates of pH in Accelerated Molecular-Tissue Aging." *Townsend Letter for Doctors* (May 1995): 34–39.

☐ Yanick, P" Functional Disturbances in Inner Ear Disorders. *Townsend Letter for Doctors* (Aug/ Sept1994): 860–863.

☐ Yanick, P. "Chronic Fatigue Syndrome & Immunosuppression." *Townsend Letter for Doctors* (April 1994): 288–290.

☐ Yanick, P. "Bioenergetic Regulation and Resiliency." *Explore* Vol. 4.5 (1993): 20–24.

☐ Yanick, P. "Lymphatic Therapy for Chronic Immune & Metabolic Disorders, Detoxification and Successful Pain Management." *Townsend Letter for Doctors* (January 1995): 34–36.

☐ Yanick, P. "MCS: Understanding Causitive Factors." *Townsend Letter for Doctors & Patients* (January 2001).

☐ Yanick, P. "New Perspectives on Allergies & Seasonal Disorders." *Townsend Letter for Doctors & Patients* (May 2001).

☐ Yanick P 2003. The Quantum Repatterning Technique *American Chiropractic,* 50.

☐ Yanick, P. *Biological Energetic Regulation Method*. Coldbrook, VT: Biological Energetic Press, 1992.

☐ Yanick, P. "Immune System Protection against Bioterrorism." *Townsend Letter for Doctors & Patients*, Dec., 2001.

☐ Yanick, P. "Novel Anti-viral Strategies." *Townsend Letter for Doctors & Patients*, Feb./March, 2002.

☐ Yanick, P 2002. Mycotoxicosis: A new emerging co-factor in Alzheimer's, environmental illness and treatment-resistant syndromes, *Townsend Letter for Doctors & Patients*, 154-6.

☐ Yanick P 2002. Oral Chelation of the Biliary Tract in CV Disease. *Townsend Letter for Doctors and Patients,* 52-55.

OTHER REFERENCES on Inner Physician and Nutrition:

☐ Tegmark, M, Wheeler. "100 Years of Quantum Mysteries." Scientific American (September 2001): 68–75.

☐ Popp, A. F. New Avenues in Medicine in Bioresonance and Multiresonance Therapy. H Brugemann, Ed. Brussels, Belgium: Haug, 1990.

☐ Fiedler, N., et al. "A Controlled Comparison of Multiple Chemical Sensitivities and Chronic Fatigue Syndrome." Psychosomatic Medicine Vol. 58 (1996): 38–49.

☐ Gruber, A. J, J. I. Hudson, and H. G. Pope, Jr. "The Management of Treatment-Resistant Depression in Disorders on the Interface of Psychiatry and Medicine: Fibromyalgia, Chronic Fatigue Syndrome, Migraine, Irritable Bowel Syndrome, Atypical Facial Pain, and Premenstrual Dysphoric Disorder." Psychiatric Clinicians of North America. Vol. 19 (1996): 351–69.

- ☐ Hudson, J. I., and H. G. Pope, Jr. "Fibromyalgia and Psychopathology: Is Fibromyalgia a Form of 'Affective Spectrum Disorder'?" Journal of Rheumatology Supplement Vol. 19 (1989): 15–22.
- ☐ Hudson, J. I., et al. "Fibromyalgia and Major Affective Disorder: A Controlled Phenomenology and Family History Study." American Journal of Psychiatry Vol. 142 (1985): 441–6.
- ☐ Walker E. A., et al. "Psychiatric Illness and Irritable Bowel Syndrome: A Comparison with Inflammatory Bowel Disease." American Journal of Psychiatry Vol. 147 (1990): 1656–61.
- ☐ Simon, G. E, et al. "Immunologic, Psychological and Neuropsychological Factors in Multiple Chemical Sensitivity: A Controlled Study." Annals of Internal Medicine Vol. 119 (1997): 97–103.
- ☐ Wood, G. C., R. P. Bentall, M. Gopfert, and R. H. Edwards. "A Comparative Psychiatric Assessment of Patients with Chronic Fatigue Syndrome and Muscle Disease." Psychological Medicine Vol. 21 (1991):619–28.
- ☐ Balk, R. A. "Severe Sepsis and Septic Shock." Critical Care Clinician Vol. 16 (2000):179–192.
- ☐ Sands K. E., et al. "Epidemiology of Sepsis Syndrome in Eight Academic Medical Centers." JAMA Vol. 278 (1997):234–240.

- ☐ Macleod, R. L. et al. "Inhibition of Intestinal Secretion by Rice." Lancet. Vol. 346 (1995): 90–92.
- ☐ Gates, J. R., et al. "Association of Dietary Factors and Selected Plasma Variables with Sex-Hormone-Binding Globulin in Rural Chinese Women." American Journal of Clinical Nutrition Vol. 63.1 (1996): 22–31
- ☐ Blobel, G., et al. "Metal Ion Chaperone Function of the Soluble Cu(I) Receptor Axis." Science. Vol. 278 (1997):853–56.
- ☐ Gushleff, B. W. " The Role of Novel Phyto-Estrogen and Progestogen Therapy in the Menopausal Patient." Informedica 314 (1986): 205.
- ☐ Saltzman, J. R., et al. "Nutritional Consequences of Intestinal Bacterial Overgrowth." Complete Therapy Vol. 20.9 (1994):523–530.
- ☐ Gebbers, J. O., et al: Immunological Structures and Functions of the Gut." Schweiz Arch Tierheilk. Vol. 131 (1989): 221–238.
- ☐ Kulli, P., et al. "Food Intolerance and Rheumatoid Arthritis." Lancet (1988): 1419–1420.
- ☐ O'Farrelly, C., et al. "Association between Villous Atrophy in Arthritis, Rheumatoid Factor...IgG." Lancet (1988): 819–822.
- ☐ Shanahan, F. A. "Gut Reaction: Lymphoepithelial Communication in the Intestine." Science Vol. 275 (1997):1897–1898.

☐ Thedorou V., et al. "Integrative Neuroimmunology of the Digestive Tract." Vet Res Vol. 27 (1996): 427–442.

☐ Wallace, J. L., et al. "Inflammatory Mediators in GI Defense and Injury." PSEBM. Vol. 214 (1997):192–203.

☐ Fiocchi, C. "Cytokines and Intestinal Inflammation." Transplant Processes Vol. 18 (1996): 394–400.

☐ Adibi, S., and E. Phillips. "Evidence for Greater Absorption of Amino Acids from Peptide Than from Free Form in Human Intestine." Clinical Research Vol. 16 (1968): 446.

☐ Craft, I. et al: "Absorption and Malabsorption of Glycine and Glycine Peptides in Man." Gut Vol 9 (1968) :425–437.

☐ Adibi, S.A et al: "Comparison of Free Amino Acid and Dipeptide Absorption in Jejunum of Sprue Patients." Gastroenterology Vol. 67 (1974): 586–591.

☐ Reicht, G., W. Petritsch, A. Eherer, et al. "Jejunal Protein Absorption of Whey Protein and Its Hydrolysate." JPEN Vol. 16 (1992): 25S.

☐ Neredith J. W., J. A. Ditesheim, and G. P. Zaloga. "Visceral Protein Levels in Trauma Patients Are Greater with Peptide Diet Than Intact Protein Diet." J Trauma Vol. 30 (1990):825–829.

☐ Gardner, M.G. "Intestinal Assimilation of Intact Peptides/Proteins from the Diet, A Neglected Field." Biol Rev Vol. 59 (1984):289–331.

☐ Boullin, D.J., R. F. Crampton, C. E. Heading , et al. Intestinal Absorption of Dipeptides Containing Glycine, Phenylalanine, Proline, B-alanine, or Histidine in Rat. Clinical Science Molecular Medicine Vol. 45 (1973):849–858.

☐ Gardner, M.G. "Absorption of Intact Peptides: Studies on Transport of Protein Digest and Dipeptides across Rat Small Intestine in Vitro." Q J Exp Physiol Vol. 67 (1982):629–637.

☐ Kontessis, P., S. Jones, R. Dodds, et al. "Renal, Metabolic and Hormonal Responses to Ingestion of Animal and Vegetable Proteins." Kidney Int Vol. 38 (1990):136–144.

☐ Silk, D.B.A., P. D. Fairclough, M. L. Clark, et al. "Use of a Peptide Rather Than Free Amino Nitrogen Source in Chemically Defined 'Elemental' Diets." JPEN Vol. 4 (1980):548–553.

☐ Keohane P.P., G. K. Grimble, B. Brown, et al. "Influence of Protein Composition and Hydrolysis Method on Intestinal Absorption of Protein in Man." Gut Vol. 26 (1985):907–913.

☐ Webb, K.E. "Amino Acid and Peptide Absorption from the Gastrointestinal Tract." Federation Proceeding Vol. 45(1986): 2268–2271.

☐ Amoss, M., J. Rivier, and R. Guillemin. "Release of Gonadotropins by Oral Administration of Synthetic LRF or a Tripeptide Fragment of LRF." J Clin Endocrinol Metab 35:175–177, 1972.

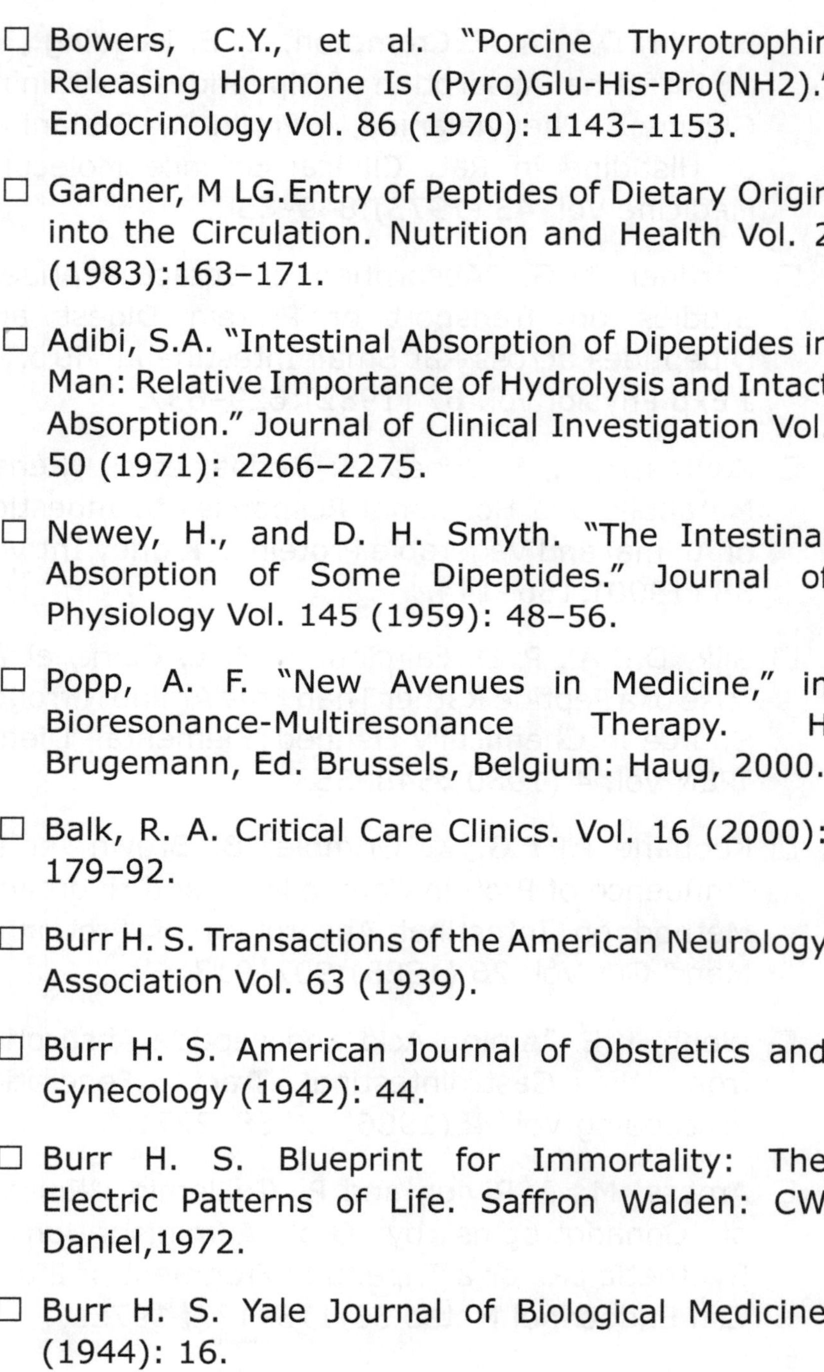

- ☐ Bowers, C.Y., et al. "Porcine Thyrotrophin Releasing Hormone Is (Pyro)Glu-His-Pro(NH2)." Endocrinology Vol. 86 (1970): 1143–1153.
- ☐ Gardner, M LG.Entry of Peptides of Dietary Origin into the Circulation. Nutrition and Health Vol. 2 (1983):163–171.
- ☐ Adibi, S.A. "Intestinal Absorption of Dipeptides in Man: Relative Importance of Hydrolysis and Intact Absorption." Journal of Clinical Investigation Vol. 50 (1971): 2266–2275.
- ☐ Newey, H., and D. H. Smyth. "The Intestinal Absorption of Some Dipeptides." Journal of Physiology Vol. 145 (1959): 48–56.
- ☐ Popp, A. F. "New Avenues in Medicine," in Bioresonance-Multiresonance Therapy. H Brugemann, Ed. Brussels, Belgium: Haug, 2000.
- ☐ Balk, R. A. Critical Care Clinics. Vol. 16 (2000): 179–92.
- ☐ Burr H. S. Transactions of the American Neurology Association Vol. 63 (1939).
- ☐ Burr H. S. American Journal of Obstretics and Gynecology (1942): 44.
- ☐ Burr H. S. Blueprint for Immortality: The Electric Patterns of Life. Saffron Walden: CW Daniel,1972.
- ☐ Burr H. S. Yale Journal of Biological Medicine (1944): 16.

- ☐ Burr H. S. Proceedings of National Academy of Science (1946): 32.
- ☐ Burr H. S. Yale Journal of Biological Medicine (1945): 17.
- ☐ Burr H. S. Federal Proceedings (1947): 6.
- ☐ Burr H. S. Yale Journal of Biological Medicine (1942): 14.
- ☐ Burr H. S. Yale Journal of Biological Medicine, (1947):19.
- ☐ Burr H. S. Yale Journal of Biological Medicine (1949): 21.
- ☐ Abrams A New Concepts in Diagnosis and Treatment. San Francisco: Philopolis Press, 1916.
- ☐ Boyd, W. Royal Society of Medicine, 1925.
- ☐ Becker, R. O. Cross Currents: The Perils of Electropollution. Putnam. NY: Tarcher, 1990.
- ☐ Lakhovsky, G. The Secret of Life. Sussex, England: True Health Publishing, 1951.
- ☐ Crile, G. The Phenomena of Life: A Radio-electric Interpretation. 1936.
- ☐ Lewis, T. British Medical Journal Vol. 431 (1937).
- ☐ Hunt, V. Progress Report: A Study of Structural Integration from Neuromuscular Energy Field and Emotional Approaches. Los Angeles, CA: UCLA, 1977.

☐ Kirlian, S. Kirlian Photography. Russia., 1949

☐ Mandell, P. Energy Emission Analysis. EV Publishing, 1988.

☐ DeVernejourl, P. The Kirilian Aura. New York: Doubleday, 1974.

☐ DeVernejourl, P. "The Kirilian Question." Bulletin of the Academy of National Medicine (1985).

☐ Kim, P. Design for Destiny. New York, NY: Ballantine Books, 1974.

☐ Tiller, W. Energy Field Observations. 1988.

☐ Motoyama, H. Science and the Evolution of Consciousness. Autumn Press, 1978.

☐ Voll, R. American Acupuncture Vol. 8. (1980).

☐ Smith, J. The Dimensions of Healing: A Symposium The Academy of Parapsychology and Medicine, 1972.

☐ Hunt, V. Science of Mind, 1982.

☐ Reynonds, R. J. "The Gas Between the Stars." Scientific American (2001): 34–43.

☐ Ferriere, K. M. in Reviews in Modern Physics Vol. 73 (2001): 4.

☐ Stoicheff, H. et al "Mitochondrial DNA and Disease." New England Journal of Medicine Vol. 334.4 (1996): 270–271.

☐ Wallace, D. C. "Mitochondrial DNA in Aging and Disease." Scientific American (1997): 40–47.

- ☐ Richter, C. "Oxidative Damage to Mitochondrial DNA and Its Relationship to Aging. Int J Biochem Cell Biol Vol. 27.7 (1995): 647–653.
- ☐ Blaylock, R. L. A "Review of Conventional Cancer Prevention and Treatment and the Adjunctive Use of Nutraceutical Supplements/Antioxidants. Is There Danger or Significant Benefit?" JAMA Vol. 3.3 (2000): 17–35.
- ☐ Loft, S., et al. "Cancer Risk and Oxidative DNA Damage in Man." J Mol Med. Vol. 74 (1996): 297–312.
- ☐ Wei, Q., et al. "DNA Repair: A Potential Marker for Cancer Susceptibility." Cancer Bulletin Vol. 46 (1994): 233–37.
- ☐ Legerski, R. J., et al. "DNA Repair Capability and Cancer Risk." Cancer Bulletin Vol. 46 (1994): 228–32.
- ☐ Noroozi, M., et al. "Effects of Flavonoids and Vitamin C on Oxidative DNA Damage to Human Lymphocytes." American Journal of Clinical Nutrition Vol. 67 (1998): 1210–18.
- ☐ Wei, Y. H., and S. H. Dao. "Mitochondrial DNA Mutations and Lipid Peroxidation in Human Aging." In C. D. Berdainer and J. L. Hargrove, Nutrients and Gene Expression. Boca Raton: CRC Press, 1996.
- ☐ Rucker, R, and D. Tinker. "The Role of Nutrition in Gene Expression: A Fertile Field for the Application

of Molecular Biology." Journal of Nutrition Vol. 116 (1986): 177–189.

☐ DeVernejourl, P. The Kirilian Aura. New York: Doubleday, 1974.

☐ Zimmerman, J. Psychosomatic Medicine 1979:11.

☐ Hunt, V. Brain Mind Bulletin Vol. 3 (1978): 9.

☐ Popp, F. A., et al. "Biophotonic Emission of the Human Body." J Photochem & Photobiol, Vol. 40 (1997): 187–89.

☐ Popp, F. A., and J. J. Chang. "Mechanism of Interaction between Electromagnetic Fields and Living Systems." Science in China Vol. 43 (2000): 507–18.

☐ Rettemyer, M., A. F. Popp, and W. Nagl. Naturwissenchaften Vol. 11 (1981): 572–3.

☐ Benveniste, J. Letter. The Lancet (1998): 351.

☐ Benveniste, J., et al. "A Simple Fast Method for In Vivo Demonstration of Electromagnetic Molecular Signaling (EMD) via High Dilution or Computer Recording." FASEB Journal Vol. 13 (1999): A163.

☐ Burr, H. S. The Fields of Life. New York: Ballantine, 1972.

☐ Gerber, R. Vibrational Medicine. Santa Fe: Bear and Company, 1988.

☐ Jensen, B., and M. Anderson. Empty Harvest. Garden City Park, NY: Avery, 1990.

☐ "Veggie Nutrients Dip in Tests." Omaha World-Herald January 29, 2000: 6.

☐ Finely, J., et al. "Selenium Content of Foods Purchased in North Dakota." Nutr. Res. Vol. 16 (1996): 723–28.

☐ World Cancer Research Fund. Food, Nutrition and the Prevention of Cancer. A Global Perspective. Washington, DC: American Institute for Cancer Research,1997.

☐ See, D. Journal of the American Nutraceutical Association Vol 2(1999): 25–41.

☐ Block, J. B., and S. Evans. "Clinical Evidence Supporting Cancer Risk Reduction with Antioxidants and Implications for Diets and Supplements. JAMA. Vol. 3.3 (2000): 6–16.

☐ Heber, D. What Color is Your Diet? New York: HarperCollins, 2001.

☐ Houston, M.D., and J. S. Strupp. "Prevention and Treatment of Cancer: Is the Cure in the Produce Aisle?" JAMA. Vol. 3.3 (2000): 27–30.

☐ Reed, M. J., and A. Purohit. "Breast Cancer and the Role of Cytokines in Regulating Estrogen Synthesis." Endocrine Review Vol. 18 (1997): 701–715.

☐ Stoicheff, H., and C. Vital. "Mitochondrial DNA and Disease." N Eng J Med Vol. 334.4 (1996): 270–271.

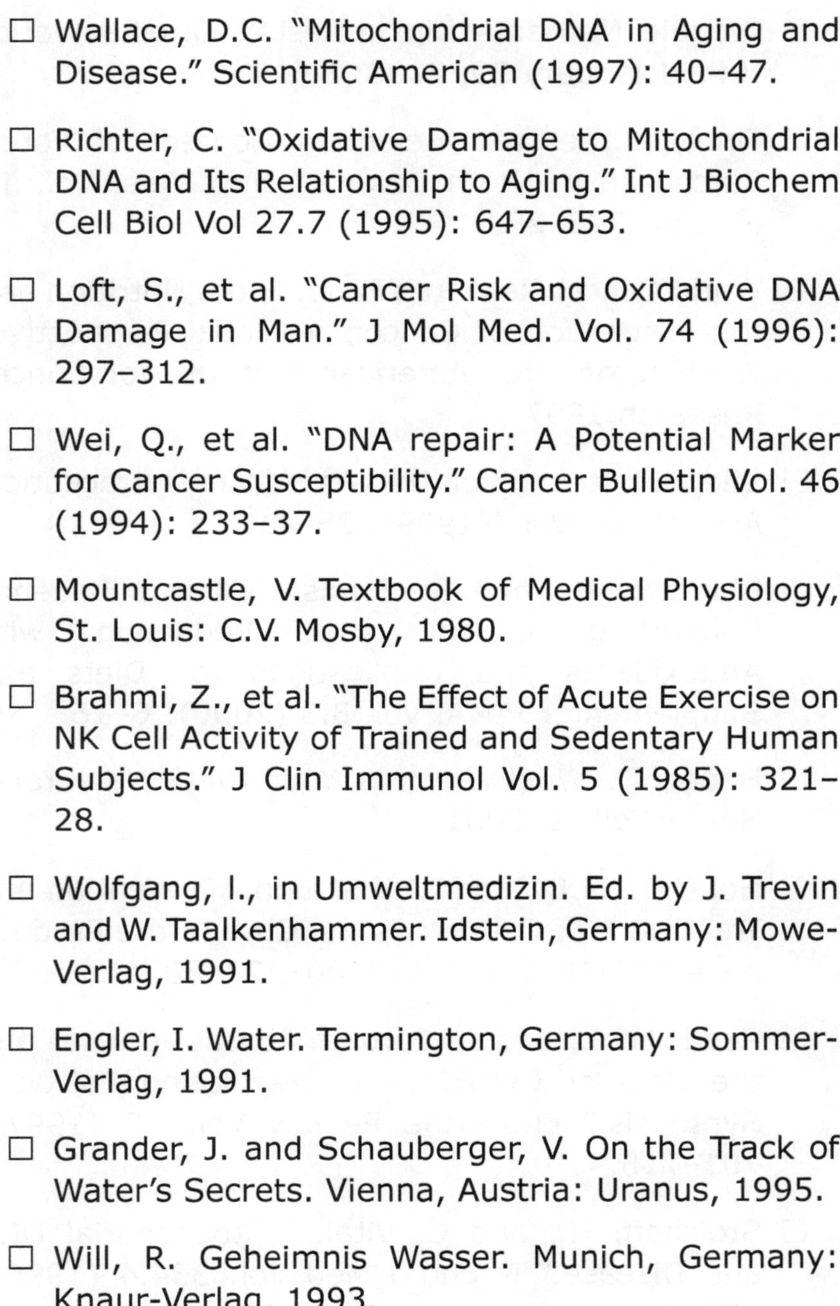

- ☐ Wallace, D.C. "Mitochondrial DNA in Aging and Disease." Scientific American (1997): 40–47.
- ☐ Richter, C. "Oxidative Damage to Mitochondrial DNA and Its Relationship to Aging." Int J Biochem Cell Biol Vol 27.7 (1995): 647–653.
- ☐ Loft, S., et al. "Cancer Risk and Oxidative DNA Damage in Man." J Mol Med. Vol. 74 (1996): 297–312.
- ☐ Wei, Q., et al. "DNA repair: A Potential Marker for Cancer Susceptibility." Cancer Bulletin Vol. 46 (1994): 233–37.
- ☐ Mountcastle, V. Textbook of Medical Physiology, St. Louis: C.V. Mosby, 1980.
- ☐ Brahmi, Z., et al. "The Effect of Acute Exercise on NK Cell Activity of Trained and Sedentary Human Subjects." J Clin Immunol Vol. 5 (1985): 321–28.
- ☐ Wolfgang, I., in Umweltmedizin. Ed. by J. Trevin and W. Taalkenhammer. Idstein, Germany: Mowe-Verlag, 1991.
- ☐ Engler, I. Water. Termington, Germany: Sommer-Verlag, 1991.
- ☐ Grander, J. and Schauberger, V. On the Track of Water's Secrets. Vienna, Austria: Uranus, 1995.
- ☐ Will, R. Geheimnis Wasser. Munich, Germany: Knaur-Verlag, 1993.

- ☐ Coburn, T., et al. Our Stolen Future. New York, NY: Penguin, 1996.
- ☐ Labrie, C, A. Belanger, and F. Labrie. "Androgenic Activity of Dehydroepiandrosterone and Androstenedione in the Rat Ventral Prostate". Endocrinology Vol. 123 (1988): 1412–1417.
- ☐ Labrie F. "Intracrinology." Mol Cell Endocrinol. Vol. 78 (1991): C113–C118.
- ☐ Roy, R., and A. Belanger. "Lipoproteins: Carriers of Dehydroepiandrosterone Fatty Acid Esters in Human Serum." J Steroid Biochem Vol. 34 (1989): 559–561.
- ☐ Provencher, P. H., R. Roy, and A. Belanger. "Pregnenolone Fatty Acid Esters Incorporated into Lipoproteins: Substrates in Adrenal Steroidogenesis." Endocrinology Vol. 130 (1992): 2717–2724.
- ☐ Roy, R., and A. Belanger. "Formation of Lipoidal Steroids in Follicular Fluid." J Steroid Biochem Vol. 33 (1989): 257–262.
- ☐ Labrie, F., A. Dupont, and A. Belanger. "Complete Androgen Blockade for the Treatment of Prostate Cancer." In: V. T. De Vita, S. Hellman, and S. A. Rosenberg, Eds. Important Advances in Oncology. Philadelphia, Lippincott: 1985.
- ☐ Brochu, M, et al. "Effects of Flutamide and Aminoglutethimide on Plasma 5a-reduced Steroid Glucuronide Concentrations in Castrated Patients

with Cancer of the Prostate." J Steroid Biochem Vol. 28 (1987): 619–622.

☐ Belanger, A., et al. "Steroid Glucuronides: Human Circulatory Levels and Formation by LNCaP Cells." J Steroid Biochem Mol Biol Vol.40 (1989): 593–598 .

☐ Labrie, F., et al. "Structure, Function and Tissue-specific Gene Expression of 3b-hydroxysteroid Dehydrogenase/5-ene-4-ene Isomerase Enzymes in Classical and Peripheral Intracrine Steroidogenic Tissues." J Steroid Biochem Mol Biol Vol. 43 (1992): 805–826.

☐ Wolfe, M. S., et al. "Blood Levels of Organochlorine Residues and Risk of Breast Cancer." J Natl Cancer Inst Vol. 85.8 (1993): 468–652.

☐ Coburn, T., et al. Our Stolen Future. New York: Penguin, 1996.

☐ Coburn, T., et al. Our Stolen Future. New York: Penguin, 1996.

☐ Stoicheff, H., and C. Vital. "Mitochondrial DNA and Disease." N Eng J Med Vol. 196 (1996): 270–271.

☐ Wallace, D. C. "Mitochondrial DNA in Aging and Disease." Scientific American (1997): 40–47.

☐ Richter, C. "Oxidative Damage to Mitochondrial DNA and Its Relationship to Aging." Int J Biochem Cell Biol Vol. 27.7 (1995): 647–653.

☐ Loft, S., et al. "Cancer Risk and Oxidative DNA Damage in Man." J Mol Med. Vol. 74 (1996): 297–312.

☐ Wei, Q., et al. "DNA Repair: A Potential Marker for Cancer Susceptibility." Cancer Bulletin. Vol. 46 (1994): 233–37.

☐ Legerski, R. J., et al. "DNA Repair Capability and Cancer Risk." Cancer Bulletin. Vol. 46 (1994): 228–32.

☐ Noroozi, M., et al "Effects of Flavonoids and Vitamin C on Oxidative DNA Damage to Human Lymphocytes." Am J Clin Nutr. Vol. 67 (1998): 1210–18.

☐ Wei, Y. H., and S. H. Dao. "Mitochondrial DNA Mutations and Lipid Peroxidation in Human Aging." In C. D. Berdainer and J. L. Hargrove. Nutrients and Gene Expression. Boca Raton, FL: CRC Press, 1996.

☐ Rucker, R., and D. Tinker. "The Role of Nutrition in Gene Expression: A Fertile Field for the Application of Molecular Biology." J Nutr Vol. 116 (1986): 177–189.

☐ Becker, R. O. Cross Currents: The Perils of Electropollution. Putnam, NY: Tarcher, 1990.

☐ Gillette, B. "Raising the Alarm: Concerns Linger About EMFs." E Magazine, Nov-Dec (2001): 40–41.

☐ Jensen, B, and M. Anderson. Empty Harvest. Garden City Park, New York: Avery, 1990.

☐ "Veggie Nutrients Dip in Tests." Omaha World-Herald, January 29 (2000): 6.

☐ Finely, J. et al: "Selenium Content of Foods Purchased in North Dakota." Nutr. Res. Vol. 16 (1996): 723–728.

☐ World Cancer Research Fund. Food, Nutrition and the Prevention of Cancer. A Global Perspective. Washington, D.C.: American Institute for Cancer Research, 1997.

☐ Hu, J., et al. "Risk Factors for Oesophageal Cancer in Northeast China." Int J Cancer Vol. 57 (1994): 38–46.

☐ Haung, M. T., et al. "Inhibition of Skin Tumorgenesis by Rosemary and its Constituents Carnosol and Ursolic Acid." Cancer Res. Vol. 54 (1994): 701–8.

☐ Steinmetz KA et al. "Vegetables, Fruit, and Cancer Prevention: A Review." J Am Diet Assoc. Vol. 96 (1996): 27–37.

☐ Winter, J., et al. Chemicals in the Human Food Chain. New York: Van Nostrand Reinhold, 1990.

☐ World Health Organization. Report on the Panel on Food and Agriculture. Geneva, Switzerland: World Health Organization, 1992.

☐ Minchin, R. F., et al. "Role of Acetylation in Colorectal Cancer." Mutat Res. Vol. 290 (1993): 35–42.

- ☐ Sinha, R., et al. "High Concentrations of the Carcinogen 2-amino-1-methel-6-phenykunudazi Occur in Chicken but Are Dependent on the Cooking Method." Cancer Res. Vol. 55 (1996): 16–19.
- ☐ Snyderwibe, E. G. "Some Perspective on the Nutritional Aspects of Breast Cancer Research. Food Derived HCAs as Etiologic Agents in Human Mammary Cancer." Cancer Vol. 74 (1994): 977–94.
- ☐ Sacarello, H. L. A. Handbook of Hazardous Materials. Washington, DC: 1994. Lewis Publishers, 1994.
- ☐ Heber, D. What Color is Your Diet? New York: HarperCollins, 2001.
- ☐ Houston, M. D., and J. S. Strupp. "Prevention and Treatment of Cancer: Is the Cure in the Produce Aisle?" JAMA Vol. 3.3 (2000): 27–30.
- ☐ Block, J. B., and S. Evans. "Clinical Evidence Supporting Cancer Risk Reduction with Antioxidants and Implications for Diets and Supplements." JAMA. Vol. 3.3 (2000): 6–16.
- ☐ Auernhammer, C.J., et al. "Effect of Growth Hormone and Insulinlike Growth Factor I on the Immune System." Eur. J. Endocrinology. Vol. 133 (1995): 635–45.
- ☐ Rudman, D; et. al. "Effect of Human Growth Hormone in Men over 60 Years Old." N. Eng. J. Med. Vol. 323 (1990):1–9.

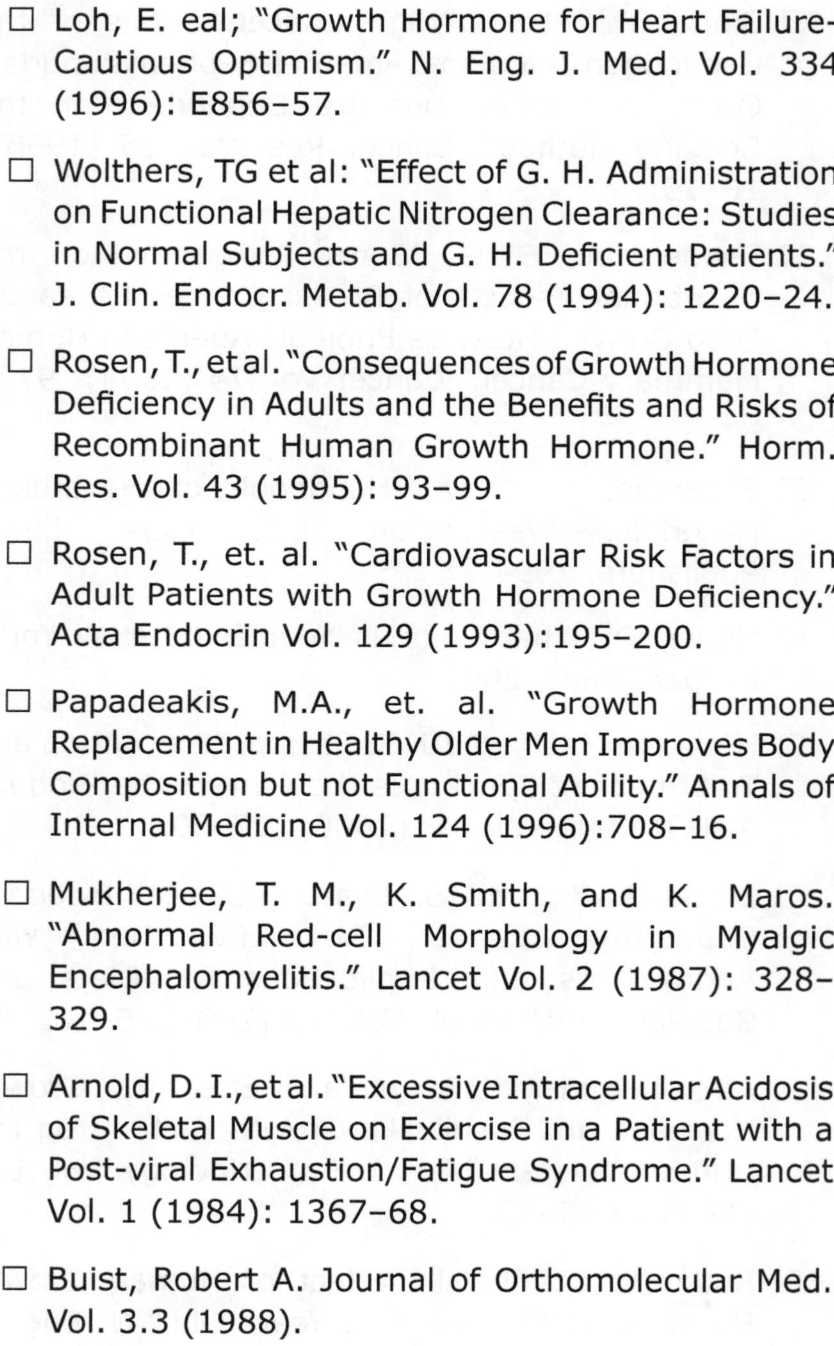

☐ Loh, E. eal; "Growth Hormone for Heart Failure-Cautious Optimism." N. Eng. J. Med. Vol. 334 (1996): E856–57.

☐ Wolthers, TG et al: "Effect of G. H. Administration on Functional Hepatic Nitrogen Clearance: Studies in Normal Subjects and G. H. Deficient Patients." J. Clin. Endocr. Metab. Vol. 78 (1994): 1220–24.

☐ Rosen, T., et al. "Consequences of Growth Hormone Deficiency in Adults and the Benefits and Risks of Recombinant Human Growth Hormone." Horm. Res. Vol. 43 (1995): 93–99.

☐ Rosen, T., et. al. "Cardiovascular Risk Factors in Adult Patients with Growth Hormone Deficiency." Acta Endocrin Vol. 129 (1993):195–200.

☐ Papadeakis, M.A., et. al. "Growth Hormone Replacement in Healthy Older Men Improves Body Composition but not Functional Ability." Annals of Internal Medicine Vol. 124 (1996):708–16.

☐ Mukherjee, T. M., K. Smith, and K. Maros. "Abnormal Red-cell Morphology in Myalgic Encephalomyelitis." Lancet Vol. 2 (1987): 328–329.

☐ Arnold, D. I., et al. "Excessive Intracellular Acidosis of Skeletal Muscle on Exercise in a Patient with a Post-viral Exhaustion/Fatigue Syndrome." Lancet Vol. 1 (1984): 1367–68.

☐ Buist, Robert A. Journal of Orthomolecular Med. Vol. 3.3 (1988).

☐ Behan, P. O., W. M. H. Behan, and E. J. Bell. "The Post-viral Fatigue Syndrome-Analysis of Findings in Fifty Cases." Journal of Infections Vol. 10 (1985): 211–222.

☐ Cox, I.M., et al. "RBC Magnesium and Chronic Fatigue Syndrome." Lancet Vol. 337 (1991): March 30.

☐ Perger, F. "Klinik der Lambliasis Intestinalis und Ihre Verbreitung in Mittleleuropa." Nautramed Vol. 3, 1988.

☐ Galland, L. "Leaky Gut Syndromes." Townsend Letter for Doctors. Aug/Sept, 1995.

☐ Masoro, E. J. "Food Restriction in Rodents: An Evaluation of Its Role in the Study of Aging." Journal of Gerontology. Vol. 43 (1988): B59–B64.

☐ Gates, J. Cornell University Study. Manuscript in Publication, 2002.

☐ Wyatt, J., et al. "International Permeability and the Prediction of Release in Crohn's Disease." Lancet Vol. 341 (1993): 1437–9.

☐ Gulbins, E., and F. Lang. "Pathogens, Host-Cell Invasion and Disease." American Scientist Vo. 89 (2001): 406–12.

☐ Ploegh, Hidde. "Viral Strategies of Immune Evasion." Science Vol. 280 (1998): 248–253.

☐ Jones PS "Strategies for Antiviral Drug Discovery." Antiviral Chem & Chemotherapy Vol. 9.4 (1998): 283–302.

☐ Bernstein, Jack M. Antiviral Chemotherapy: General Overview. Dayton, OH: Wright State University School of Medicine, Division of Infectious Diseases, 2000.

☐ Page, Roderic D. M., and Edward C. Holmes. Molecular Evolution. Boston, MA: Blackwell Science, 1988.

☐ Collier, Leslie, and John Oxford. Human Virology. Oxford, England: Oxford University Press, 2000.

☐ Cody, V., et al. Plant Flavonoids in Biology and Medicine. Biochemical, Cellular, and Molecular Properties New York: Alan R. Liss, 1988.

☐ Borchers, A. T., et al. "Mushrooms, Tumors, and Immunity." Proceedings of the Society of Experimental Biological Medicine Vol. 4 (1999): 282–93.

☐ Kabara, J. J., et al. "Fatty Acids and Derivatives as Antimicrobial Agents." Antimicrobial Agents and Chemotherapy (1992): 23–28.

☐ Kabara, J. J. "Toxicological, Bacteriocidal and Fungicidal Properties of Fatty Acids and Some Derivatives." JAOCS Vol. 56 (1979): 760.

☐ Hierholzer, J. C., and Kabara, J. J. "In Vitro Effects of Monolaurin Compounds on Enveloped RNA and DNA Viruses." J. Food Safety Vol. 4 (1982): 1–12.

☐ Enig, M. G. Coconut Oil: An Anti-bacterial, Anti-viral Ingredient for Food, Nutrition and Health. Manila, Philippines: AVOC Luric Symposium, 1997.

☐ McTaggert, L. The Field: The Quest for the Secret Force of the Universe. New York: Harper Collins, 2001.

☐ Popp, A. F. Biophotonen. Heidelberg, Germany: Schriftenreihe Krebsgeschehen, 1984.

☐ Verastegui, M. Angeles, et al. "Antimicrobial Activity of Extracts of Three Major Plants of the Chichuahuan Desert." J of Enthopharmacology Vol. 52 (1996): 175–77.

☐ Brinker, F. "Larrea tridentata." British J of Phytotherapy Vol. 3.1. (1994): 10–30.

☐ Kawagishi, H., et al. "Herinacines A,B,C, Strong Stimulators of Nerve Growth Factor from the Mushroom Hericium." Tetrahedron Letters Vol. 32 (1991): 4561–64.

☐ Kawagishi, H., et al: "Herinacines A,B,C, Strong Stimulators of Nerve Growth Factor from the Mushroom Hericium." Tetrahedron Letters Vol. 35.10 (1994): 1569–72.

☐ Scott, D. W., and W. L. Scott. The Extremely Unfortunate Skull Valley Incident. Chelmsford, Canada: Chelmsford Publishers, 2001.

☐ Scott, D. W., and W. L. Scott. The Brucellosis Triangle. Chelmsford, Canada: Chemsford Publishers, 1998.

☐ Lo, S. "Pathogenic Mycoplasma." US Patent 5,242,820, issued 9–7–93.

☐ World Health Organization. Report on the Panel on Food and Agriculture. Geneva, Switzerland: World Health Organization, 1992.

☐ Bell, I. R "A Time-Dependent Sensitization in Environmental Illness: A Pharmacologic Model." The Eleventh International Symposium on Man and His Environment in Health and Disease.1993. November 29, 2002 http://www.aehf.com/articles/1993symp.html.

☐ Bell, I. R., et al. "An Olfactory-limbic Model of Multiple Chemical Sensitivity Syndrome." Biological Psychiatry Vol. 32 (1992): 218–242.

☐ Lorig, T., et al. "EEG Activity during Administration of Low-concentration Odors." Bulletin of Psychonomic Science Vol. 28 (1990): 405–8.

☐ Gilbert, M. E. "Neurotoxicants and Limbic Kindling." In R. L. Isaacs and K. F. Jensen, Eds., The Vulnerable Brain and Environmental Risks, Vol. I. New York: Plenum Press.1992.

☐ Constantini, A. V. " The Fungal/Mycotoxin Etiology of Atherosclerosis and Hyperlipidemia." 1993. The Eleventh International Symposium on Man and His Environment in Health and Disease. November 29, 2002 http://www.aehf.com/articles/1993symp.html.

☐ Richard, J. L. "In G. E. Bray and D. H. Ryan, Eds. Mycotoxins, Cancer, and Health. Baton-Rouge, LA: Louisiana State University Press, 1991.

☐ Delincee, H., and B. Pool-Aobel. "Genotoxic Properties of 2–dodecycobutanone, a Compound Formed by Irradiation of Food Containing Fat." Radiation Physics and Chemistry Vol. 52 (1988): 39–42.

☐ Le Tellier, P. R., and W. W. Mawar. "2–alkalcyclobutanones from the Radiolysis of Triglycerides." Lipids Vol. 7 (1972): 76–76.

☐ Delincee, H., et al. "Genotoxicity of 2–alkalcyclobutanones Markers for Irradiation Treatment in Fat-Containing Food. A paper presented at the 12th International Meeting on Radiation Processing. Avignon, France: March 25–30, 2001.

☐ Miller, C. S. "Toxicology and Industrial Health." New Engl. J. Med. 335.2A (1992):1498–1504.

☐ Enig, M. G. "Coconut, in Support of Good Health in the 21st Century." 36th Asian Pacific Coconut Community, 1999.

☐ Enig, M. G. Coronary Heart Disease: The Dietary Sense and Nonsense. London, England: Janus Publishing, 1993.

☐ Fife, B. The Miracles of Coconut Oil. Colorado Springs, CO: Healthwise. 2000.

☐ Kabara. J. J., et al. "Fatty Acids and Derivatives as Antimicrobial Agents." Antimicrobial Agents and Chemotherapy (1992): 23–28.

☐ Kabara. J. J. "Toxicological, Bacteriocidal and Fungicidal Properties of Fatty Acids and Some Derivatives." JAOCS Vol. 56 (1979): 760.

☐ Hierholzer, J. C., and Kabra J. J. "In Vitro Effects of Monolaurin Compounds on Enveloped RNA and DNA Viruses. J. Food Safety Vol 4 (1982): 1–12.

☐ Enig, M. G. Coconut Oil: An Anti-bacterial, Anti-viral Ingredient for Food, Nutrition and Health. Manila, Philippines: AVOC Luric Symposium, 1997.

☐ Loviselli, A., et al. "Low Levels of Dehydroepiandrosterone Sulfate in Adult Males with Insulin-Dependent Diabetes Mellitus. Minerva Endocrinology Vol. 19 (1994): 113–119.

☐ Van Vollenhoven, R. F., et al. "An Open Study of Dehydroepiandrosterone in Systemic Lupus Erthematosus." Arthritis Rheum Vol. 37 (1994):1305–1310.

☐ Jacobson, M., et al. "Decreased Serum Dehydroepiandrosterone Is Associated with an Increased Progression of Human Immunodeficiency Virus Infection in Men with CD4 Cell Counts of 200–499." Journal of Infectious Diseases, Vol. 164.5 (1991): 864 (5).

☐ Ebeling, P., et al. "Physiological Importance of Dehydroepiandrosterone." The Lancet Vol. 343 (1994): 1479.

☐ Fava, M., et al. "Dehydroepiandrosterone-Sulfate/ Cortisol Ratio in Panic Disorder. Psychiatry Res. Vol. 28 (1989): 345–350.

☐ Atschule, M., and J. Kitay. McLead Hospital Pineal Research Collections, 1940.

☐ Beck-Friis, J., et al. "Serum Melatonin in Relation to Clinical Variables in Patients with Major Depressive Disorder and a Hypothesis of a Low Melatonin Syndrome." Acta Psychiatrica Scandinavia Vol. 71(1985): 319–30.

☐ Cavallo, A., et al. "Melatonin Circadian Rhythm in Childhood Depression." Journal of the American Academy of Child and Adolescent Psychiatry Vol. 26.3 (1987): 395–99.

☐ Selkoe, D.J. "Amyloid Protein and Alzheimer's Disease." Scientific American Vol. 4 (1993): 54–58.

☐ Linder, M.E., and A. G. Gilman. "G Proteins." Scientific American, 1993.

☐ Pischinger, A. Matrix and Matrix Regulation. Brussels, Belgium: Haug Publishers, 1991.

☐ Kartner, N., and V. Ling. "Multidrug Resistance in Cancer." Scientific American, 1993.

☐ Carskadon, M.A., and C. Acebo. "Parental Reports of Seasonal Mood and Behavior Changes in

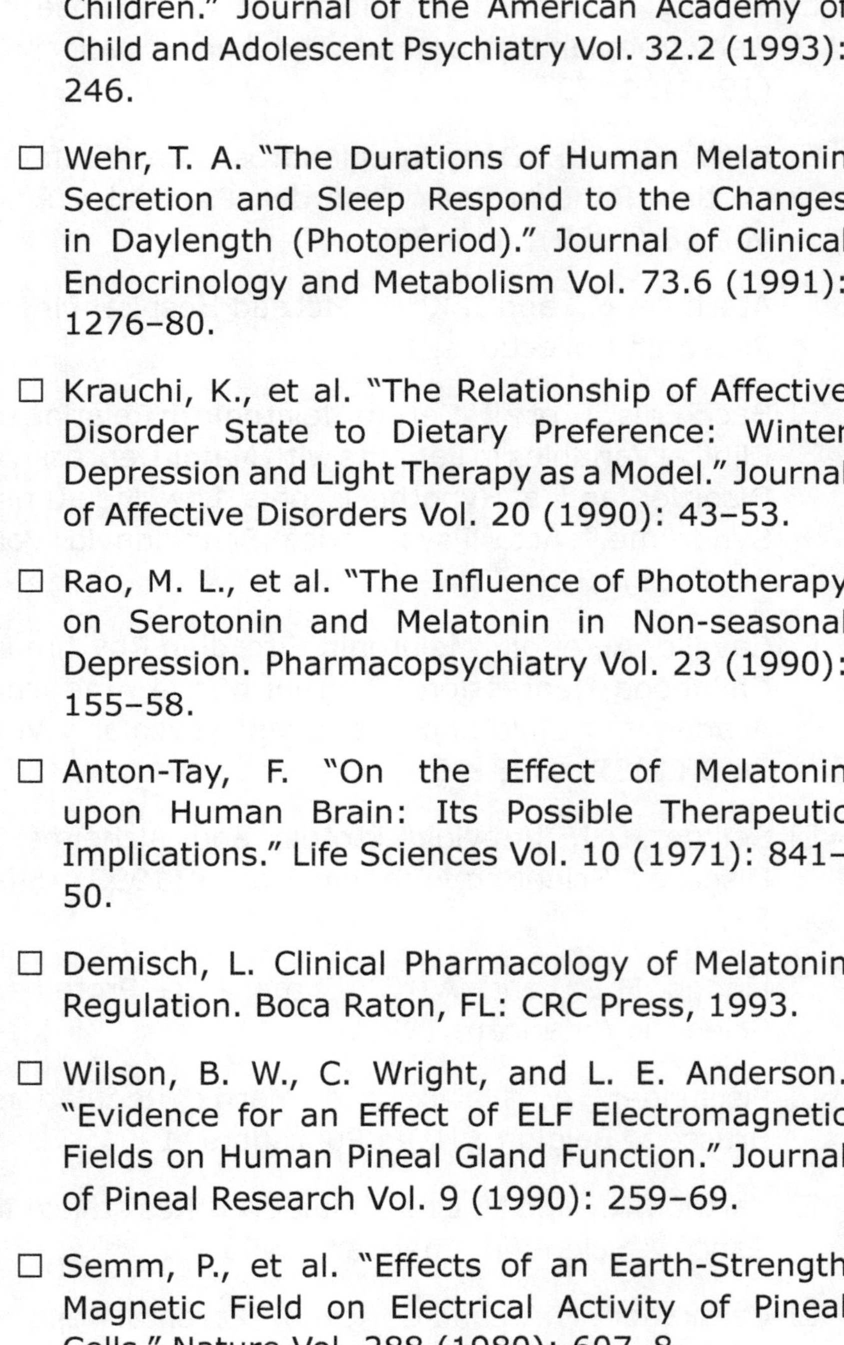

Children." Journal of the American Academy of Child and Adolescent Psychiatry Vol. 32.2 (1993): 246.

☐ Wehr, T. A. "The Durations of Human Melatonin Secretion and Sleep Respond to the Changes in Daylength (Photoperiod)." Journal of Clinical Endocrinology and Metabolism Vol. 73.6 (1991): 1276–80.

☐ Krauchi, K., et al. "The Relationship of Affective Disorder State to Dietary Preference: Winter Depression and Light Therapy as a Model." Journal of Affective Disorders Vol. 20 (1990): 43–53.

☐ Rao, M. L., et al. "The Influence of Phototherapy on Serotonin and Melatonin in Non-seasonal Depression. Pharmacopsychiatry Vol. 23 (1990): 155–58.

☐ Anton-Tay, F. "On the Effect of Melatonin upon Human Brain: Its Possible Therapeutic Implications." Life Sciences Vol. 10 (1971): 841–50.

☐ Demisch, L. Clinical Pharmacology of Melatonin Regulation. Boca Raton, FL: CRC Press, 1993.

☐ Wilson, B. W., C. Wright, and L. E. Anderson. "Evidence for an Effect of ELF Electromagnetic Fields on Human Pineal Gland Function." Journal of Pineal Research Vol. 9 (1990): 259–69.

☐ Semm, P., et al. "Effects of an Earth-Strength Magnetic Field on Electrical Activity of Pineal Cells." Nature Vol. 288 (1980): 607–8.

☐ Wilson, B. W., et al. "Neuroendocrine Mediated Effects of Electromagnetic-Field Exposure: Possible Role of the Pineal Gland." Life Sciences Vol. 45 (1985): 1319–32.

☐ "Correlation Between Heart Attacks and Magnetic Activity." Nature Vol. 277 (1994): 646–48.

☐ Dubbels, R., et al. "Melatonin Determination with a Newly Developed ELISA System: Inter-individual Differences in the Response of the Human Pineal Gland to Magnetic Fields." In G. J. Maestroni, A. Conti, and R. J. Reiter, Eds. Advances in Pineal Research, Vol. 7. London: John Libbey and Co., 1994.

☐ Doll, R., et al. "The Causes of Cancer." J Nat Cancer Institute Vol. 66: (1981): 1191.

☐ World Cancer Research Fund. "Food, Nutrition and the Prevention of Cancer. A Global Perspective." Washington, D.C.: American Institute for Cancer Research, 1997.

☐ Cummings, J. H., et al. "Diet and the Prevention of Cancer." BMJ Vol. 317 (1998): 1636–40.

☐ Houston, M. D., and J. S. Strupp. "Prevention and Treatment of Cancer: Is the Cure in the Produce Aisle?" JAMA Vol. 3.3 (2000): 27–30.

☐ Block J. B., and S. Evans. "Clinical Evidence Supporting Cancer Risk Reduction with Antioxidants and Implications for Diets and Supplements." JAMA. Vol. 3.3 (2000): 6–16.

☐ Jensen, B., and M. Anderson. Empty Harvest. Garden City Park, New York.: Avery, 1990.

☐ "Veggie Nutrients Dip in Tests." Omaha World-Herald, January 29,2000: 6.

☐ Finely, J., et al. "Selenium Content of Foods Purchased in North Dakota." Nutr. Res. Vol. 16 (1996): 723–28.

☐ Hu, J., et al. "Risk Factors for Oesophageal Cancer in Northeast China." Int J Cancer Vol. 57 (1994): 38–46.

☐ Haung, M. T., et al. "Inhibition of Skin Tumorgenesis by Rosemary and Its Constituents Carnosol and Ursolic Acid." Cancer Research Vol. 54 (1994): 701–8.

☐ Steinmetz, K. A., et al. "Vegetables, Fruit, and Cancer Prevention: A Review." J Am Diet Association Vol. 96 (1996): 27–37.

☐ Winter, et al. Chemicals in the Human Food Chain. Basel, Germany: Van Nostrand Reinhold, 1990.

☐ World Health Organization. Report on the Panel on Food and Agriculture. Geneva, Switzerland: World Health Organization, 1992.

☐ Minchin, R. F., et al. Role of Acetylation in Colorectal Cancer. Mutation Research Vol. 290 (1993.): 35–42.

☐ Sinha, R., et al. "High Concentrations of the Carcinogen 2–amino-1–methel-6–phenykunudazi Occur in Chicken but Are Dependent on the

Cooking Method." Cancer Res Vol. 55 (1996): 4516–19.

- ☐ Snyderwibe, E. G. "Some Perspective on the Nutritional Aspects of Breast Cancer Research. Food Derived HCAs as Etiologic Agents in Human Mammary Cancer." Cancer Vol. 74 (1994): 977–94.
- ☐ Sacarello, H. L. A. Handbook of Hazardous Materials. Washington, DC: Lewis Publishers, 1994.
- ☐ See, D. Journal of the American Nutraceutical Association Vol. 2.1 (1996): 25–41.
- ☐ Marshall, R. Ten Secrets You May Not Know. Round Rock, TX: Premier Research Labs, 2001.
- ☐ Khachik, F., et al. "Distribution, Bioavailability and Metabolism of Carotenoids in Humans." In W. R. Bidlack, S. T. Omaye, et al, Eds. Phytochemicals: A New Paradigm. Basel, Germany: Technomic Pub, Inc., 1998.
- ☐ Kuhlmann, M. K., et al. "Reduction of Cisplatinin Toxicity in Cultured Renal Tubular Cells by the Bioflavonoid Quercitin." Arch Toxicology Vol. 72 (1998): 536–40.
- ☐ Venkatesan, N. "Curcumin Attenuation of Acute Adriamycin Myocardial Toxicity in Rats." Br. J Pharm. Vol. 124 (1998): 425–27.
- ☐ Furr, H. C., et al. "Intestinal Absorption and Tissue Distribution of Carotenoids." Nutr Biochemistry Vol. 8 (1997): 364–77.

☐ Ferriola, P C et al. "Protein Kinase C Inhibition by Plant Flavonoids." Biochem Pharmacol. Vol. 38 (1989): 1617–24.

☐ Agullo, G., et al. "Relationship between Structure and Inhibition of Phosphotidylinositol 3-kinase: a Comparison with Tyrosine Kinase and Protein Kinase C Inhibition." Biochem pharmacology Vol. 53 (1997): 1649–57.

☐ Hoffman, J., et al. "Enhancement of the Antiproliferative Effect of Cis-damminedichloroplatinum (II) and Nitrogen Mustard by Inhibitors of Protein Kinase C." Int J Cancer Vol. 42 (1998): 382–88.

☐ Scambia, G., et al. "Inhibitory Effect of Quercitin on Primary Ovarian and Endometrial Cancers and Synergistic Activity with Cis-damminedichloroplatinum (II)." Gyn Oncology Vol.45 (1992): 13–19.

☐ Wang, Z., et al. "Mammary Cancer Promotion and MAPK Activation Associated with Consumption of a Corn Oil-based High Fat Diet." Nutr Cancer Vol. 34: (1999): 140–46.

☐ Hurston, S. D., et al. "Types of Dietary Fat and the Incidence of Cancer at Five Sites." Prev Med. Vol. 9 (1990): 242–53.

☐ Vojdani, A., et al. "New Evidence for the Antioxidant Properties of Vitamin C." Intern Soc of Prev Oncology Vol. 24.6 (2000): 508–523.

☐ Vojdani, A., and G. Namatalla. "Enhancement of NK Cytoxic Activity by Vitamin C in Pure and Augmented Formulations." J Nutr Envirn Medicine Vol. 7 (1997): 187–95.

☐ Havlick, H. D. "Functional Foods: Science or Marketing?" JAMA Vol. 4.1 (2001): 9–10.

☐ Macleod, R. L., et al. "Inhibition of Intestinal Secretion by Rice." Lancet (1995): 90–92.

☐ Gates, J. R., et al. "Association of Dietary Factors and Selected Plasma Variables with Sex-hormone-binding Globulin in Rural Chinese Women." Am J Clin Nutrition Vol. 36 (1996): 22–31.

☐ Blobel G et al. "Metal Ion Chaperone Function of the Soluble Cu(I) Receptor Axis." Science 278 (1997): 853–56.

☐ Gushleff, B. W. "The Role of Novel Phyto-Estrogen and Progestogen Therapy in the Menopausal Patient." Informedica Vol. 314 (1986): 205.

☐ Adibi, S., E. Phillips. "Evidence for Greater Absorption of Amino Acids from Peptide than from Free Form in Human Intestine." Clin Research Vol. 16 (1968): 446.

☐ Craft, I L et al. "Absorption and Malabsorption of Glycine-Glycine Peptides in Man." Gut Vol. 9 (1968): 425–437.

☐ Adibi, SA et al; "Comparison of Free Amino Acid and Dipeptide Absorption in the Jejunum of Sprue Patients." Gastroenterology Vol. 67 (1974): 586–591.

- ☐ Reicht, G. et al. "Jejunal Protein Absorption of Whey Protein and Its Hydrolysate." JPEN Vol. 16 (1992): 25S.

- ☐ Neredith J W et al:. "Visceral Protein Levels in Trauma Patients Are Greater with Peptide Diet than Intact Protein Diet." J Trauma Vol 30 (1990): 825–829.

- ☐ Gardner, M.G. "Intestinal Assimilation of Intact Peptides and Proteins from the Diet-A Neglected Field." Biol Review Vol. 59 (1984): 289–331.

- ☐ Boullin, DJ et al: "Intestinal Absorption of Dipeptides Containing Glycine, Phenylalanine, Proline, B-alanine, or Histidine in the Rat." Clinical Science Molecular Medicine Vol. 45 (1973): 849–858.

- ☐ Gardner, M.G. "Absorption of Intact Peptides: Studies on Transport of Protein Digest and Dipeptides across Rat Small Intestine in Vitro." Q J Exp Physiol Vol. 67 (1982): 629–637.

- ☐ Kontessis, P., S. Jones, R. Dodds, et al. "Renal, Metabolic and Hormonal Responses to Ingestion of Animal and Vegetable Proteins." Kidney Int Vol. 38 (1990): 136–144.

- ☐ Silk, D.B.A., P. D. Fairclough, M. L. Clark, et al. "Use of a Peptide Rather than Free Amino Nitrogen Source in Chemically Defined 'Elemental' Diets." JPEN Vol. 4 (1980): 548–553.

- ☐ Keohane P.P., G. K. Grimble, B. Brown, et al. "Influence of Protein Composition and Hydrolysis

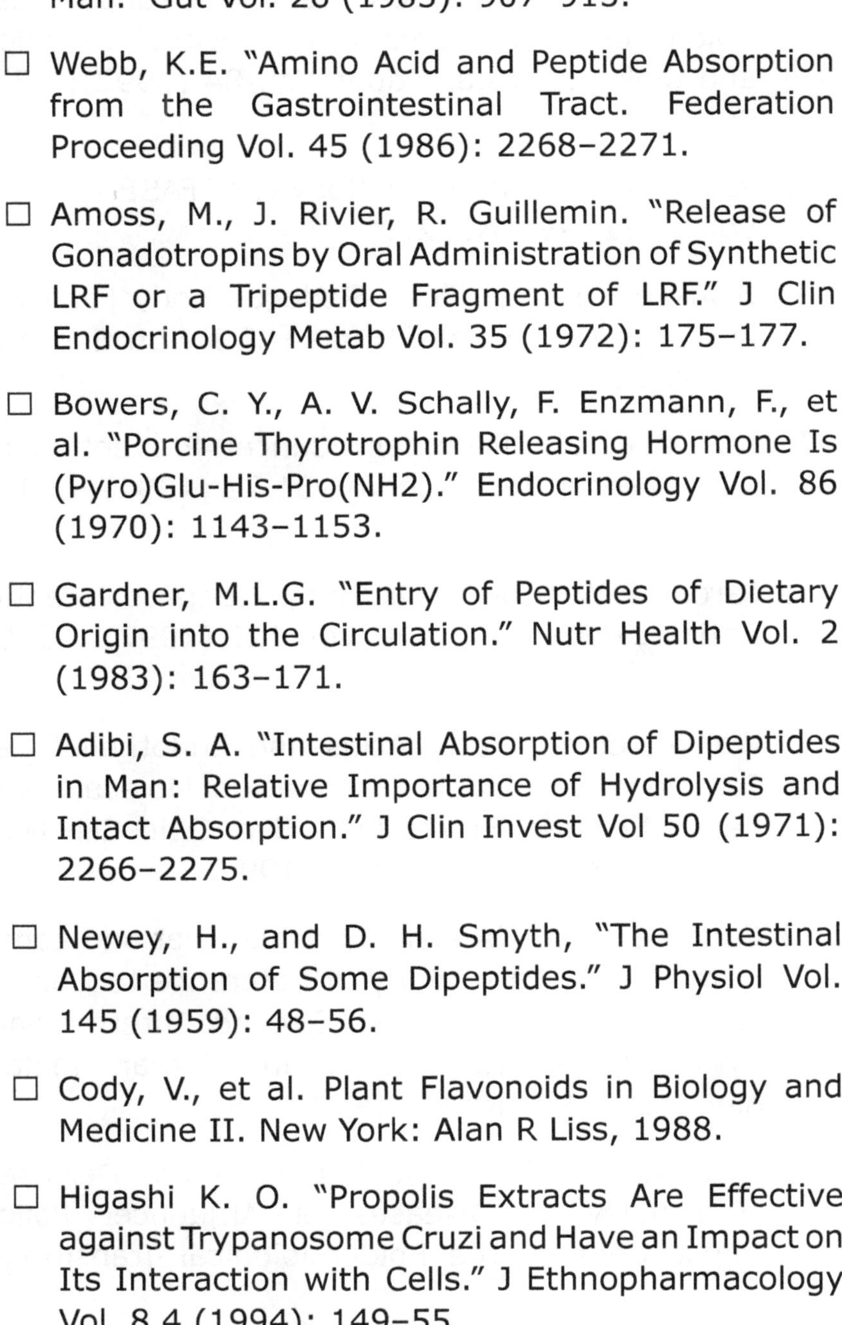

Method on Intestinal Absorption of Protein in Man." Gut Vol. 26 (1985): 907–913.

☐ Webb, K.E. "Amino Acid and Peptide Absorption from the Gastrointestinal Tract. Federation Proceeding Vol. 45 (1986): 2268–2271.

☐ Amoss, M., J. Rivier, R. Guillemin. "Release of Gonadotropins by Oral Administration of Synthetic LRF or a Tripeptide Fragment of LRF." J Clin Endocrinology Metab Vol. 35 (1972): 175–177.

☐ Bowers, C. Y., A. V. Schally, F. Enzmann, F., et al. "Porcine Thyrotrophin Releasing Hormone Is (Pyro)Glu-His-Pro(NH2)." Endocrinology Vol. 86 (1970): 1143–1153.

☐ Gardner, M.L.G. "Entry of Peptides of Dietary Origin into the Circulation." Nutr Health Vol. 2 (1983): 163–171.

☐ Adibi, S. A. "Intestinal Absorption of Dipeptides in Man: Relative Importance of Hydrolysis and Intact Absorption." J Clin Invest Vol 50 (1971): 2266–2275.

☐ Newey, H., and D. H. Smyth, "The Intestinal Absorption of Some Dipeptides." J Physiol Vol. 145 (1959): 48–56.

☐ Cody, V., et al. Plant Flavonoids in Biology and Medicine II. New York: Alan R Liss, 1988.

☐ Higashi K. O. "Propolis Extracts Are Effective against Trypanosome Cruzi and Have an Impact on Its Interaction with Cells." J Ethnopharmacology Vol. 8.4 (1994): 149–55.

☐ Rao, C. V., et al. "Effect of Caffeic Acid Esters on Carcinogen-induced Mutagenecity and Human Colon Adenocarcinoma Cell Growth." Chemical and Biological Interactions Vol. 84 (1992): 277–90.

☐ Pariza, M. "CLA Reduces Body Fat." FASEB Journal Vol. 10 (1996): A3227.

☐ Blankson H et al. "CLA Reduces Body Fat Mass in Obese/Overweight Humans." J Nutrition 130 (2000): 2943–8.

☐ Ip, C., et al. "Mammary Cancer Prevention by CLA." Cancer Research Vol. 51.22 (1991): 6118–24.

☐ Parodi, P. W. "Cow's Milk Components as Potential Anticarcinogenic." J Nutr Vol. 127 (1997): 1055–60.

☐ Wahke, et al. "Fatty Acids and Endothelial Cell Function: Regulation of Adhesion Molecule and Redox Enzyme Expression." Curr Opin Clin Nutr Metab Care Vol. 2.2 (1999): 109–15.

☐ Campbell, T. C. "A Plant-Enriched Diet and Long-term Health, Particularly in Reference to China." Paper presented at the Second International Symposium on Horticulture and Human Health, Alexandria, VA: November 4, 1989.

☐ Campbell, T. C. et al., "China: From Diseases of Poverty to Diseases of Affluence. Policy Implications of the Epidemiological Transition."

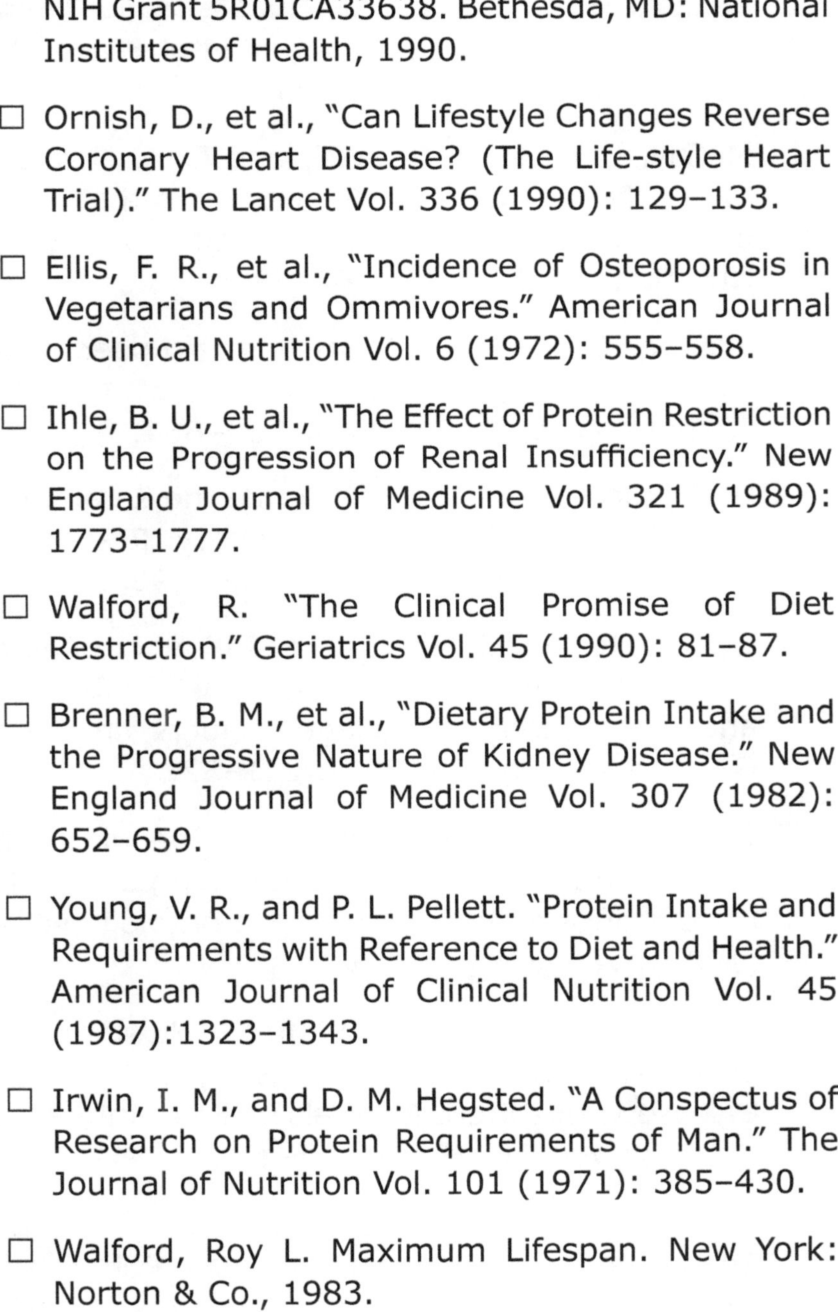

NIH Grant 5R01CA33638. Bethesda, MD: National Institutes of Health, 1990.

☐ Ornish, D., et al., "Can Lifestyle Changes Reverse Coronary Heart Disease? (The Life-style Heart Trial)." The Lancet Vol. 336 (1990): 129–133.

☐ Ellis, F. R., et al., "Incidence of Osteoporosis in Vegetarians and Ommivores." American Journal of Clinical Nutrition Vol. 6 (1972): 555–558.

☐ Ihle, B. U., et al., "The Effect of Protein Restriction on the Progression of Renal Insufficiency." New England Journal of Medicine Vol. 321 (1989): 1773–1777.

☐ Walford, R. "The Clinical Promise of Diet Restriction." Geriatrics Vol. 45 (1990): 81–87.

☐ Brenner, B. M., et al., "Dietary Protein Intake and the Progressive Nature of Kidney Disease." New England Journal of Medicine Vol. 307 (1982): 652–659.

☐ Young, V. R., and P. L. Pellett. "Protein Intake and Requirements with Reference to Diet and Health." American Journal of Clinical Nutrition Vol. 45 (1987):1323–1343.

☐ Irwin, I. M., and D. M. Hegsted. "A Conspectus of Research on Protein Requirements of Man." The Journal of Nutrition Vol. 101 (1971): 385–430.

☐ Walford, Roy L. Maximum Lifespan. New York: Norton & Co., 1983.

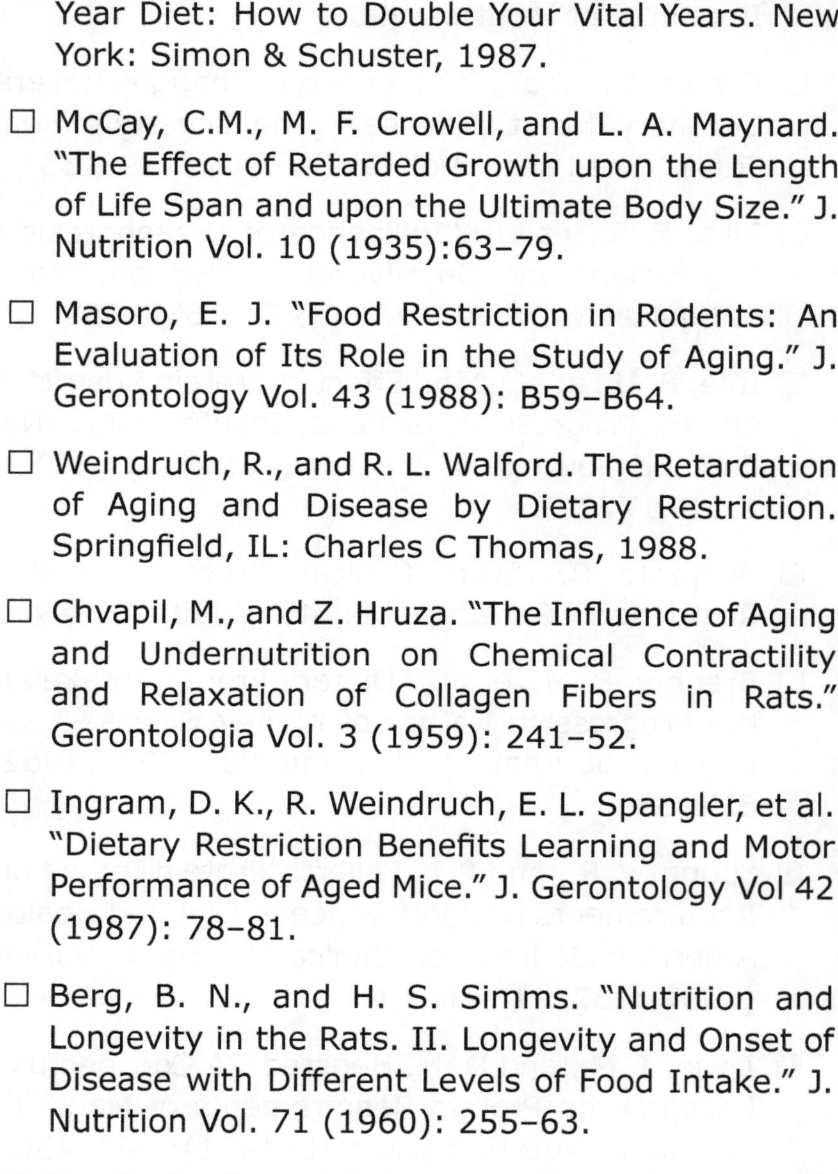

- ☐ Walford, Roy L. The One Hundred and Twenty Year Diet: How to Double Your Vital Years. New York: Simon & Schuster, 1987.
- ☐ McCay, C.M., M. F. Crowell, and L. A. Maynard. "The Effect of Retarded Growth upon the Length of Life Span and upon the Ultimate Body Size." J. Nutrition Vol. 10 (1935):63–79.
- ☐ Masoro, E. J. "Food Restriction in Rodents: An Evaluation of Its Role in the Study of Aging." J. Gerontology Vol. 43 (1988): B59–B64.
- ☐ Weindruch, R., and R. L. Walford. The Retardation of Aging and Disease by Dietary Restriction. Springfield, IL: Charles C Thomas, 1988.
- ☐ Chvapil, M., and Z. Hruza. "The Influence of Aging and Undernutrition on Chemical Contractility and Relaxation of Collagen Fibers in Rats." Gerontologia Vol. 3 (1959): 241–52.
- ☐ Ingram, D. K., R. Weindruch, E. L. Spangler, et al. "Dietary Restriction Benefits Learning and Motor Performance of Aged Mice." J. Gerontology Vol 42 (1987): 78–81.
- ☐ Berg, B. N., and H. S. Simms. "Nutrition and Longevity in the Rats. II. Longevity and Onset of Disease with Different Levels of Food Intake." J. Nutrition Vol. 71 (1960): 255–63.
- ☐ Yu, B. P., E. J. Masoro, I. Murata, et al. "Life Span Study of SPF Fischer: 344 Male Rats Fed Ad Libitum or Restricted Diets: Longevity, Growth,

Lean Body Mass and Disease." J Gerontology Vol. 37 (1982):130–141.

☐ McCay, Clive M., et al. "The Life Span of Rats on a Restricted Diet." Journal of Nutrition Vol. 18 (1939):1–25.

☐ McCay, Clive M. "Effect of Restricted Feeding upon Aging and Chronic Disease in Rats and Dogs." American Journal of Public Health Vol. 37 (1947): 521.

☐ Mazess, R. B., and W. Mather. "Bone Mineral Content of North Alaskan Eskimos." American Journal of Clinical Nutrition Vol. 27 (1974): 916–925.

☐ Williams, Clyde. "Diet and Endurance Fitness." American Journal of Clinical Nutrition Vol. 49 (1989): 1077–1083.

☐ Masoro, E. J., I. Shimokawa, and B. P. Yu. "Retardation of Aging Process in Rats by Food Restriction." Ann. NY Acad. Sci. Vol. 621 (1991): 337–52.

☐ Yu, B. P. "Food Restriction Research: Past and Present Status." Rev. Biol. Res. Aging Vol. 4 (1990): 349–71.

☐ Jung, L K et al: "Effect of Calorie Restriction on the Production and Responsiveness to Interleukin-2 in (NZBXZW) f-1 Mice." Clinical Immunology Immunopathology Vol. 25(1982): 295–301.

☐ Perpaoli, W., A. Dall'ara, E. Pedrinis, and W. Regelson. "The Pineal Control of Aging. The

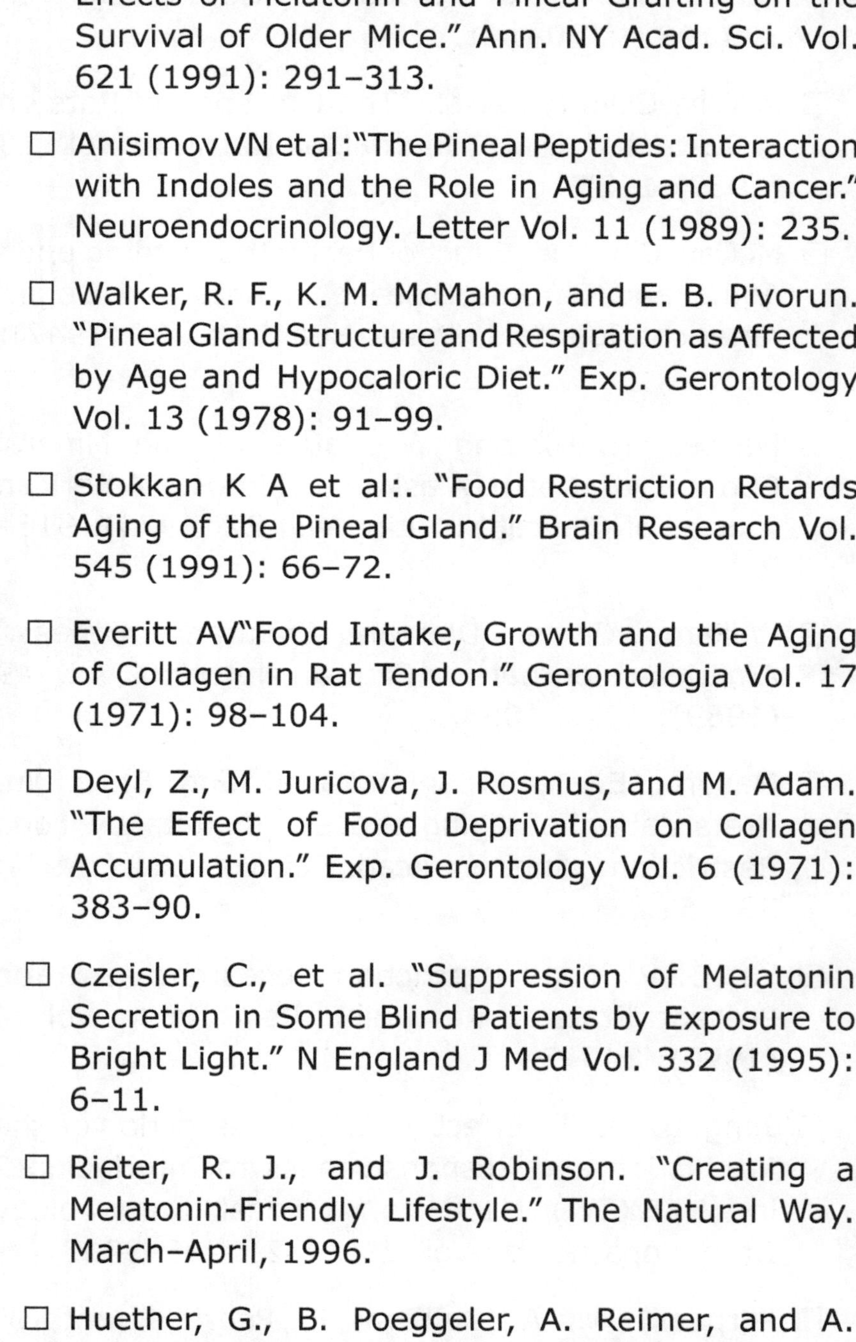

Effects of Melatonin and Pineal Grafting on the Survival of Older Mice." Ann. NY Acad. Sci. Vol. 621 (1991): 291–313.

☐ Anisimov VN et al:"The Pineal Peptides: Interaction with Indoles and the Role in Aging and Cancer." Neuroendocrinology. Letter Vol. 11 (1989): 235.

☐ Walker, R. F., K. M. McMahon, and E. B. Pivorun. "Pineal Gland Structure and Respiration as Affected by Age and Hypocaloric Diet." Exp. Gerontology Vol. 13 (1978): 91–99.

☐ Stokkan K A et al:. "Food Restriction Retards Aging of the Pineal Gland." Brain Research Vol. 545 (1991): 66–72.

☐ Everitt AV"Food Intake, Growth and the Aging of Collagen in Rat Tendon." Gerontologia Vol. 17 (1971): 98–104.

☐ Deyl, Z., M. Juricova, J. Rosmus, and M. Adam. "The Effect of Food Deprivation on Collagen Accumulation." Exp. Gerontology Vol. 6 (1971): 383–90.

☐ Czeisler, C., et al. "Suppression of Melatonin Secretion in Some Blind Patients by Exposure to Bright Light." N England J Med Vol. 332 (1995): 6–11.

☐ Rieter, R. J., and J. Robinson. "Creating a Melatonin-Friendly Lifestyle." The Natural Way. March–April, 1996.

☐ Huether, G., B. Poeggeler, A. Reimer, and A. George. "Effect of Tryptophan Administration on

Circulating Melatonin Levels in Chicks and Rats: Evidence for Stimulation of Melatonin Synthesis and Release in the Gastrointestinal Tract." Life Sciences Vol. 51 (1992): 945–953.

- ☐ Ames, B. N. (1983) Dietary carcinogens and anticarcinogens. *Science 221,* 1256-1264.

- ☐ Ames, B. N., Magaw, R. and Gold, L. S. (1987) Ranking possible carcinogenic hazards. *Science* ***236***, 271-280.

- ☐ Larone, D. H. 1995. Medically Important Fungi - A Guide to Identification, 3rd ed. ASM Press, Washington, D.C.

- ☐ Sutton, DA et al:(ed.). 1998. *Guide to Clinically Significant Fungi*, 1st ed. Williams & Wilkins, Baltimore.

- ☐ de Hoog, G S et al: 2000. *Atlas of Clinical Fungi*, 2nd ed, vol. 1. Centraalbureau voor Schimmelcultures, Utrecht, The Netherlands

- ☐ Morris, JT et al: 1991. Sporobolomyces infection in an AIDS patient. *J Infect Dis*. 164:623-624.

- ☐ Plzas, J et al: 1994. Sporobolomyces salmonicolor lympadenitis in an AIDS patient. Pathogen or passenger? *AIDS*. 8:387-388.

- ☐ Bergman, AG and CA. Kauffman. 1984. Dermatitis due to Sporobolomyces infection. *Arch Derm*. 120:1059-1060

☐ Misra, VC and HS Randhawa. 1976. Sporobolomyces salmonicolor var. fischerii, a new yeast. *Arch Microbiol*. 108:141-143.

☐ Rantala, A et al: 1995. Yeasts in blood cultures: Impact of early therapy. *Scand. J. Infect. Dis*. 21:557-561.

☐ Metcalf, RL 1971. The chemistry and biology of pesticides. In: White-Stevens, R. (Ed.) *Pesticides in the Environment, Part II,* Vol. 1, Marcel Dekker, New York, pp. 1-144.

☐ Wilson, CL and Wisniewski ME 1989. Biological control of postharvest diseases of fruits and vegetables an emerging technology. *Annu. Rev. Phytopathol.* 27:425-441.

☐ Mari, M et al: 1996. Postharvest biological control of grey mould (*Botrytis cinerea* Pers.: Fr.) on fresh-market tomatoes with *Bacillus amyloliquefaciens. Crop Prot*. 15:699-705.

☐ Mari, M et al: 1996. Biological control of grey mold in pears by antagonistic bacteria. *Biol. Control* 7:30-37.

☐ Gueldner, RC et al: 1988 Isolation and identification of iturins as antifungal peptides in biological control of peach brown rot with *Bacillus subtilis. J. Agric.Food Chem.* 36:366-370.

☐ Wilson, CL et al: 1987. Biological control of *Rhizopus* rot of peach with *Enterobacter cloacae. Phytopathology* 77:303.

☐ Wisniewski, ME and Wilson, CL 1992. Biological control of postharvest diseases of fruits and vegetables: *Recent advances. HortScience* 27:94-98.

☐ Zimand, G et l: 1996. Effect of *Trichoderma harzianum* on *Botrytis cinerea* pathogenicity. *Phytopathology* 86:1255-1260.

☐ Uraz, G et al: 2000. The Comparison of Proteinase Enzyme Activities of Bacillus Species on Skim Milk Agar and YCB-BSA (Yeast Carbon Base Bovine Serum Albumine) Agar. *Biotechnology 2000. The World Congress on Biotechnology,* 3-8, Berlin, Vol: 2, p: 353-355.

☐ Uraz, G et al: 2000. Comparison of the Antagonistic Effects of Bacillus on The *Listeria monocytogenes* (680 x L. Monocytogenes 2167 type 1) Invarious Salt and Glucose Concentration on The BHI Medium. *Biotechnology 2000. The World Congress on Biotechnolog* 2000, Berlin, Vol:2, p: 350-352.

☐ Uraz, G et al: 2000. The Inhibitory Effect of Pediococcus and Leuconostoc Strains Isolated from Raw Milk Against *Listeria monocytogenes. Biotechnology 2000. The World Congress on Biotechnology* 2000, Berlin, Vol: 2, p: 339-341.

☐ Ceponl, Y et al: Ljubljana University, Biotechnical faculty, Večna pot 111, Ljubljana, Slovenia1. Public Health Reserach Institute, 225 Warren Street, Newark, New Jersey 07103, USA2.

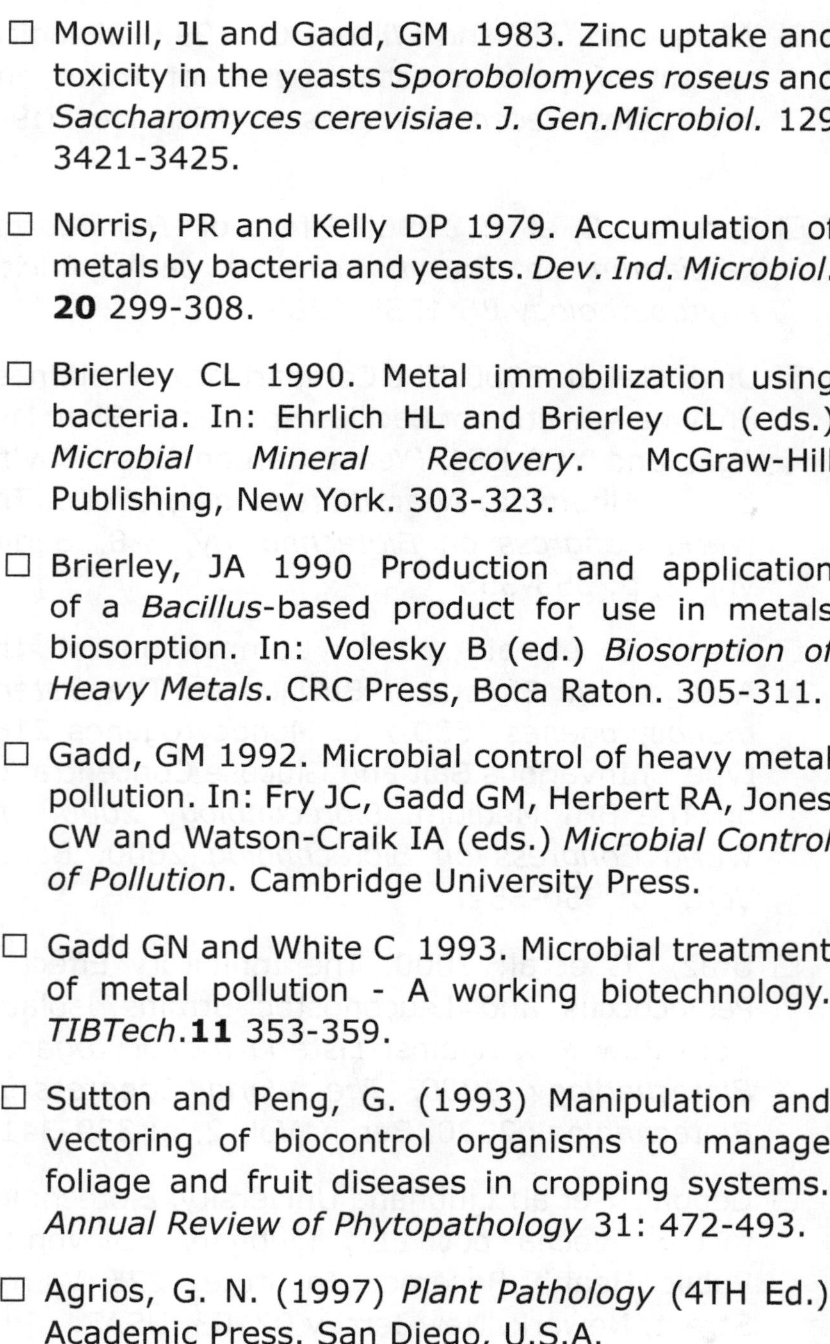

- ☐ Mowill, JL and Gadd, GM 1983. Zinc uptake and toxicity in the yeasts *Sporobolomyces roseus* and *Saccharomyces cerevisiae*. *J. Gen.Microbiol.* 129 3421-3425.
- ☐ Norris, PR and Kelly DP 1979. Accumulation of metals by bacteria and yeasts. *Dev. Ind. Microbiol.* **20** 299-308.
- ☐ Brierley CL 1990. Metal immobilization using bacteria. In: Ehrlich HL and Brierley CL (eds.) *Microbial Mineral Recovery*. McGraw-Hill Publishing, New York. 303-323.
- ☐ Brierley, JA 1990 Production and application of a *Bacillus*-based product for use in metals biosorption. In: Volesky B (ed.) *Biosorption of Heavy Metals*. CRC Press, Boca Raton. 305-311.
- ☐ Gadd, GM 1992. Microbial control of heavy metal pollution. In: Fry JC, Gadd GM, Herbert RA, Jones CW and Watson-Craik IA (eds.) *Microbial Control of Pollution*. Cambridge University Press.
- ☐ Gadd GN and White C 1993. Microbial treatment of metal pollution - A working biotechnology. *TIBTech.***11** 353-359.
- ☐ Sutton and Peng, G. (1993) Manipulation and vectoring of biocontrol organisms to manage foliage and fruit diseases in cropping systems. *Annual Review of Phytopathology* 31: 472-493.
- ☐ Agrios, G. N. (1997) *Plant Pathology* (4TH Ed.) Academic Press. San Diego, U.S.A.

- ☐ Scholberg, P.L. Marchi, A. and Bechard, J. (1995) Biocontrol of postharvest diseases of apple using *Bacillus* spp. Isolated from stored apples. *Can. J. Microbiol*. 41: 247-252.

- ☐ Swadling, I.R & Jefferies, P (1998) Antagonistic properties of two bacterial biocontrol agents of grey mould diseases. *Biocontrol Science and Technology* 8: 439-448.

- ☐ Ki-Jong Rhee et al*:* 2004.Role of Commensal Bacteria in Development of Gut-Associated Lymphoid Tissues and Preimmune Antibody Repertoire *The Journal of Immunology*, 172: 1118-1124.

- ☐ Williams, J. 2003. Portal to the interior: Viral pathogenesis and natural compounds that restore mucosal immunity and modulate inflammation. *Alternative Medicine Review*. 8(4):395-409.

- ☐ Wayne, L et al. 2000. Resistance of Bacillus Endospores to extreme Terrestrial and Extraterrestrial Environments. *Micribiol Mol Biol Rev*. 64(3): 548-72.

- ☐ Zamenoff, F et al: 1965.Genetic factors in radiation resistance of bacillus subtilis. *J of Bacteriology*. 1965, 90:180-15.

- ☐ Setlow, P. 1992. I will survive: protecting and repairing spore DNA. *J. Bacteriol*. 174:2737-2741

- ☐ Bonomo, R et al: 1980. Ruolo delle IgA secretorie nelle funzioni dell'immunita locale dell'apparato

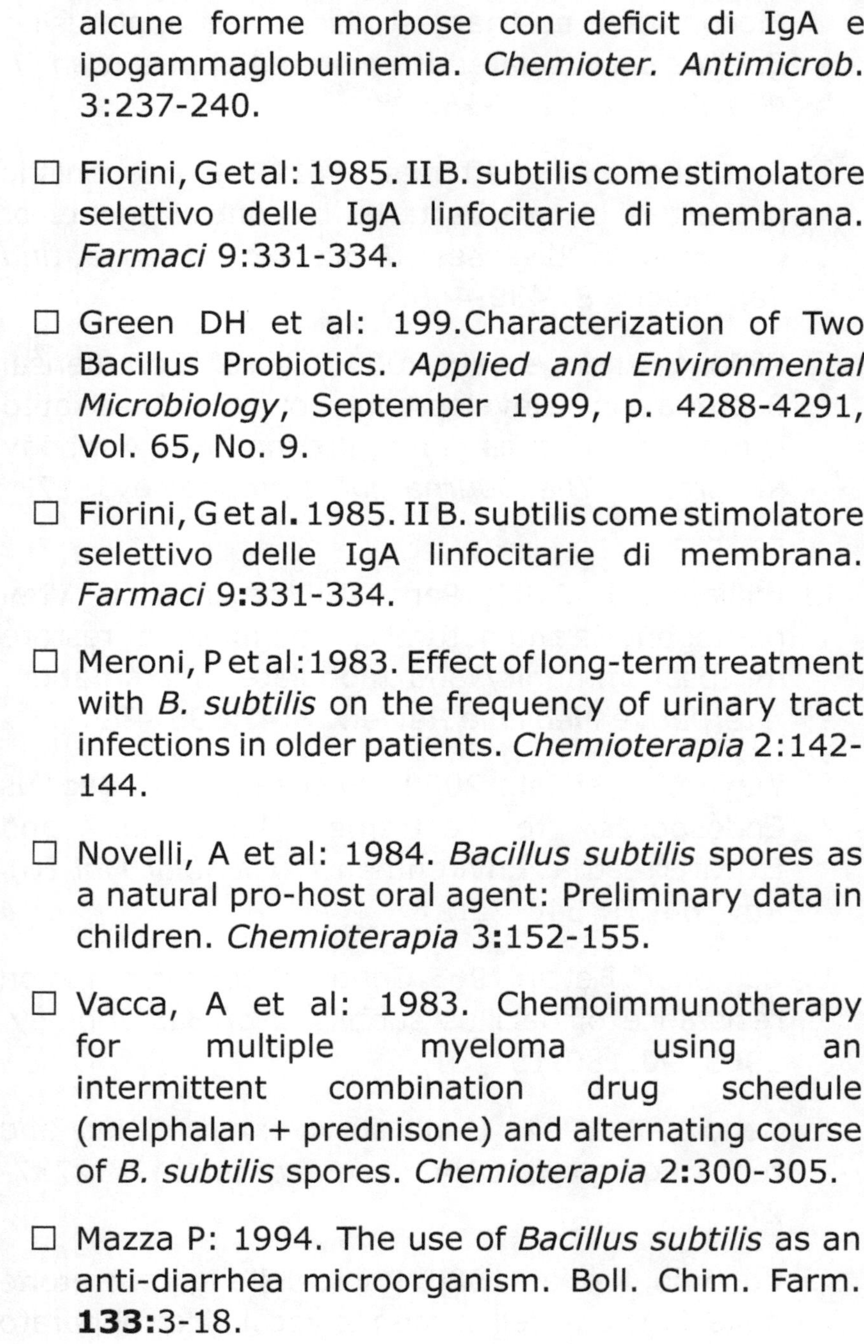

digerente. Impiego di spore di *B. subtilis* in alcune forme morbose con deficit di IgA e ipogammaglobulinemia. *Chemioter. Antimicrob.* 3:237-240.

- ☐ Fiorini, G et al: 1985. II B. subtilis come stimolatore selettivo delle IgA linfocitarie di membrana. *Farmaci* 9:331-334.
- ☐ Green DH et al: 199.Characterization of Two Bacillus Probiotics. *Applied and Environmental Microbiology*, September 1999, p. 4288-4291, Vol. 65, No. 9.
- ☐ Fiorini, G et al. 1985. II B. subtilis come stimolatore selettivo delle IgA linfocitarie di membrana. *Farmaci* 9:331-334.
- ☐ Meroni, P et al:1983. Effect of long-term treatment with *B. subtilis* on the frequency of urinary tract infections in older patients. *Chemioterapia* 2:142-144.
- ☐ Novelli, A et al: 1984. *Bacillus subtilis* spores as a natural pro-host oral agent: Preliminary data in children. *Chemioterapia* 3:152-155.
- ☐ Vacca, A et al: 1983. Chemoimmunotherapy for multiple myeloma using an intermittent combination drug schedule (melphalan + prednisone) and alternating course of *B. subtilis* spores. *Chemioterapia* 2:300-305.
- ☐ Mazza P: 1994. The use of *Bacillus subtilis* as an anti-diarrhea microorganism. Boll. Chim. Farm. **133:**3-18.

- ☐ Rowland I. Probiotics and benefits to human health— the evidence in favour. *Environ Microbiol* 1999;**1**:375–382.

- ☐ Ngo Thi Hoa: 2000. Characterization of *Bacillus* Species Used for Oral Bacteriotherapy and Bacterioprophylaxis of Gastrointestinal Disorders *Appl Environ Microbiol*. Dec; 66(12): 5241–5247.

- ☐ Casula G, Cutting SM. 2002. Bacillus probiotics: spore germination in the gastrointestinal tract. *Appl Environ Microbiol* 68:2344-52.

- ☐ Ciprandi G et al: GW. 1986. In vitro effects of B. subtilis on the immune response. *Chemioterapia* V(6):404-7.

- ☐ Vacca A et al. 1983 Chemoimmunotherapy for multiple myeloma using an intermittent combination drug schedule (Melphalan + Prednisone) and alternating courses of Bacillus subtilis spores. *Chemioterapia* II(5):300-6.

- ☐ Muscetola M et al: 1992 Effects of B. subtilis spores on interferon production. *Pharmacological Res* 26 (2): 176-7.

- ☐ Hyronimus, B et al**:** 1998. Coagulin, a bacteriocin-like inhibitory substance produced by *Bacillus coagulans*. *J. Appl. Microbio*l. 85**:**42-50.

- ☐ Naclerio G et al: 1993. Antimicrobial activity of a newly identified bacteriocin of *Bacillus cereus*. *Appl. Environ. Microbiol*. 59:4313-4316

☐ Logan, N A and P De Vos. 1998. *Bacillus* et industrie. *Bull. Soc. Fr. Microbiol*. 13:130-136.

☐ Naclerio GE et al: 1993. Antimicrobial activity of a newly identified bacteriocin of *Bacillus cereus*. *Appl. Environ. Microbiol*. 59:4313-4316

☐ Perez C et al: 1992. Production of antimicrobial by *Bacillus subtilis* MIR 15. *J. Biotechnol*. 26:331-336.

☐ Kajimura Y et al: 1996. Fusarcidin A, new depsipeptide antibiotic 2produced by *Bacillus polymyxa* KT-8. Taxonomy, fermentation, isolation, structure elucidation and biological activity. *J. Antibiot*. 49:129-135.

☐ Naclerio G et al: 1993. Antimicrobial activity of a newly identified bacteriocin of *Bacillus cereus*. *Appl. Environ. Microbiol*. 59:4313-4316.

☐ Sato, T et al: 1992. A new isocoumarin antibiotic, Y-05460M-A. *J. Antibiot*. 45:1949-1952

☐ Sorokulova IB et al: 1997. Probiotics against *Campylobacter* pathogens. *J. Travel Med*. 4:167-170.

☐ Canedi, L M et al: 1997. PM-94128, a new isocoumarin antitumor agent produced by a marine bacterium. *J. Antibiot*. 50:175-176.

☐ Itoh, JT et al: 1982. Isolation, physicochemical properties and biological activities of amicoumacins produced by *Bacillus pumilus*. *Agric. Biol. Chem*. 46:1255-1259.

☐ Osipova, IG et al**:** 1998. Safety of bacteria of the genus *Bacillus*, forming the base of some probiotics. *Zh. Mikrobiol. Epidemiol. Immunobiol*. **6:**68-70.

www.ingramcontent.com/pod-product-compliance
Lightning Source LLC
Chambersburg PA
CBHW051724260326
41914CB00031B/1724/J
* 9 7 8 1 4 2 6 9 1 6 8 1 6 *